D0154347

BASIC STRUCTURAL THEORY

This book introduces the basic equations of the theory of structures. Conventional presentations of these equations follow the ideas of elastic analysis, introduced nearly two hundred years ago. The present text is written against the background of advances made in structural theory during the last fifty years, notably by the introduction of the so-called plastic theory. Tests on real structures in the twentieth century revealed that structural states predicted by elastic analysis cannot in fact be observed in practice, whereas plastic ideas can be used to give accurate estimates of strength. Strength is discussed in the first part of this book without reference to equations of elastic deformation. However, the designer is concerned also with stiffness, for which elastic analysis is needed, and the standard equations (suitable, for example, for computer programming) are presented. Finally, stability is analyzed, which again is essentially an elastic phenomenon, and it is shown that a higher factor of safety is required to guard against buckling than is required to guarantee straightforward strength. The emphasis throughout this book is on the derivation and application of the structural equations, rather than on details of their solution (nowadays best done by computer), and the numerical examples are deliberately kept simple.

Jacques Heyman is former Head of the Department of Engineering at the University of Cambridge and the author of fourteen books, including *The stone skeleton, Elements of the theory of structures, Structural analysis: a historical approach, Elements of stress analysis* and the two-volume set, *Plastic design of frames: Volume 1, Fundamentals* and *Volume 2, Applications*, with Lord Baker. He is a Fellow of the Society of Antiquaries, the Institution of Civil Engineers and the Royal Academy of Engineering, and an Honorary Fellow of the Royal Institute of British architects. He was a consulting engineer for a number of English cathedrals and a member of the Architectural Advisory Panel for Westminster Abbey and the Cathedrals Fabric Commission for England. He also has served on many British Standards committees. Professor Heyman's *The stone skeleton* won the Choice Outstanding Academic Books Award in 1996.

Books by the author

Plastic design of portal frames
Beams and framed structures
Plastic design of frames (2 volumes)
Coulomb's memoir on statics
Equilibrium of shell structures
Elements of stress analysis
The masonry arch
The stone skeleton
Elements of the theory of structures
Arches, vaults and buttresses
Structural analysis: a historical approach
The science of structural engineering

Basic Structural Theory

Jacques Heyman

University of Cambridge

CAMBRIDGE UNIVERSITY PRESS
Cambridge, New York, Melbourne, Madrid, Cape Town, Singapore, São Paulo, Delhi

Cambridge University Press
32 Avenue of the Americas, New York, NY 10013-2473, USA

www.cambridge.org
Information on this title: www.cambridge.org/9780521897945

First published 2008

Printed in the United States of America

A catalog record for this publication is available from the British Library.

Library of Congress Cataloging in Publication Data

Heyman, Jacques.
Basic structural theory / Jacques Heyman.
p. cm.
Includes bibliographical references and index.
ISBN: 978-0-521-89794-5 (hardback)
1. Structural analysis (Engineering) – Mathematical models. I. Title.
TA645.H463 2008
624.1′7–dc22 2008005287

ISBN 978-0-521-89794-5 hardback

Contents

Preface

University courses in structural theory (as in any other branch of engineering) aim to teach the principles of the subject. It is in fact difficult, if not impossible, to discuss a principle in the abstract, and students are usually engaged in carrying out an assortment of algebraic or numerical calculations for particular examples of structures, in the hope that fundamental truths will be revealed.

Structural equations are straightforward and are, with some exceptions, linear. The equations may be written easily, but unfortunately they are very numerous. Although their solution presents no conceptual difficulty, the work involved is so heavy that, before the advent of the electronic computer, it was virtually impossible to obtain exact numerical results for any but the simplest structure. Advances in structural theory in the nineteenth century, and in the twentieth, were directed on the one hand to the establishment of basic theorems to guide the engineer towards easier formulation of the equations, and on the other hand to the development of computational techniques which could lead to approximate solutions of any required degree of accuracy.

Many of these advances were made by scientists and engineers of experience and insight, and they show great creative genius. Thus the student's understanding will be enlightened by, for example, the elastic reciprocal theorems, the concepts of strain and potential energy, and by the theory underlying the testing of models. Side by side with these fundamental elastic properties, however, the student may well be presented with a host of techniques such as deflexion coefficients, slope/deflexion equations, and moment/area methods, which might seem to be basic to the theory of structures whereas, despite their intellectual power, they are really no more than aids to the solution of the structural equations. The student may well feel aggrieved to have spent time mastering methods of calculation, when the modern computer is furnished with programs which can produce numerical solutions for any complex structure.

Some of these topics are presented in Chapter 4, but the discussion is brief. It is the intention of this book to present the basic ideas of structural theory, rather than to review the many techniques of calculation for elastic structures. These basic ideas will enable the engineer to appreciate the way in which a computer program delivers its solutions, without necessarily investigating every detail of the computation. All of these analyses, the theory and the calculations, fall within the gigantic intellectual framework of the classical theory of structures, enunciated formally by Navier in 1826, and developed over the next century to the point where it forms the basis of most design codes throughout the world.

There is a second and powerful reason for not concentrating on the conventional syllabus of nineteenth and twentieth century courses on structural engineering. The methods of elastic analysis, whether executed by hand or by computer, purport to describe the actual state of a given structure under a given loading system. The Navier theory appeared to be so self-evidently correct that it was almost a century before tests on real structures revealed that the results of elastic analysis cannot be observed in practice. There is now overwhelming evidence that the state of an actual structure may be very different from that calculated confidently by the elastic designer.

A seemingly artificial example, the four-legged table, reveals the problem. The tripod is an ideal structure – the forces in the three legs which result from a given loading can be found easily and unequivocally from simple equations. However, those same equations are insufficient to furnish the leg forces for the conventional table with four legs, and the full apparatus of elastic structural theory leads to those difficult calculations to which the computer can now give a precise answer.

This computer output is the Navier elastic solution for this theoretical structure. The real table, placed on a hard floor, will rock, and if a leg is clear of the floor by a mere fraction of a millimetre, it is certain that the force in that leg is actually zero, whereas the computer has supplied a definite value for the force. Moreover, a cork wedge may be used to make the table comfortable for its users, which exposes the task facing the structural engineer: how are the leg forces to be evaluated, so that the legs may be designed, when any one of the four may be in contact with a (supposedly) rigid floor, or clear of the floor, or supported by an elastic wedge of unknown properties?

A real structure is, in fact, supported externally in a way which is unknown (and unknowable) to the engineer, who nevertheless is required to make a design. In modelling the structure for analysis, the conventional elastic designer is forced to make some assumptions (as is the computer program) – for the table, for example, that all four legs are in contact initially with a (rigid) floor. These assumptions, seemingly innocuous and actually of small consequence, can lead to structural solutions widely different from those observed in practice. Very small differences in *boundary conditions* can lead to wholly disproportionate differences in internal structural forces, in real multi-storey buildings as well as in the simple model of the table. The foundations of a steel or concrete frame can settle by small but – for the user – acceptable amounts; a bolted joint, assumed to be inflexible, may slip on first loading; frame members may be manufactured with slight dimensional errors. Such defects seem trivial, and they do not, in fact, affect the basic strength of a structure, but it is these defects which reveal that elastic calculations give a poor indication of how a structure carries its loads.

The anomaly was fully revealed in the first half of the twentieth century by tests on buildings under construction, and the results led to the development of the so-called plastic theory as an alternative to elastic analysis. Plastic theory in its simple form makes no use of (unknowable) boundary conditions in the assessment of strength – indeed, no attempt is made to calculate the actual state of a structure. Instead, new and powerful theorems – above all, the safe theorem – give the designer confidence in plastic methods. Paradoxically, it is the safe theorem which shows that conventional elastic methods, the Navier schema – while predicting a state which is not experienced by the real structure, and which will lead to a design which is usually wasteful of material – nevertheless gives a safe estimate of strength.

However, simple plastic theory is concerned only with the prediction of strength, and traditional elastic computations cannot be rejected. Even if, as is the case, the working state of a structure is essentially unknowable, in the sense that the internal forces are critically dependent on seemingly trivial unknown imperfections, the designer may be required to make estimates of stiffness to ensure that deflexions lie below specified limits. A structure may well yield on first loading, but stiffness is basically an elastic structural property, and must be estimated somehow by the engineer. Similarly, although buckling may also involve some yielding, the onset of instability can be determined (with some empirical imprecision) by the use of classical elastic differential equations.

This book starts, then, with a discussion of the strength of structures, and it will be seen that statements can be made with some confidence. Subsequent chapters discuss stiffness and stability, and some of the traditional methods of elastic analysis are presented in order to predict deflexions and the onset of buckling. The examples are deliberately kept simple, and the necessary mathematical foundations of the subject are outlined in three short appendices.

BASIC STRUCTURAL THEORY

1 Introduction

The structures discussed in this book are assemblages of elements (e.g. beams, columns, struts, ties) that form a construction of some practical use. For example, a light steel gantry may be needed to support a cable to power electric trains; or simple portal frames, steel or concrete, may house a factory; or the elements may be combined into a framework for a multi-storey building. A *theory of structures* is necessary to ensure that the design of any particular construction will be satisfactory when built.

The designer decides on the general form of a structure – for example, using girders working in bending for a small-span bridge, rather than a lattice truss with members working in tension or compression (alternative forms may be examined simultaneously to achieve a best design). Design requirements (e.g. specified imposed loads, permitted maximum deflexions) are stated, and the designer's task is to satisfy those requirements. The design process falls logically into two stages: dimensions are assigned to the members of the chosen form, and the theory of structures is then used to ensure that the members are comfortable, and that the overall behaviour of the structure meets the criteria. This process, in general, is circular; the structural analysis cannot be made until the sizes of the members have been chosen, but those sizes depend on the results of the structural analysis. In some cases it may be possible to achieve a direct design without this circular process of trial and error (and, certainly, computer programs may be written to achieve rapidly convergent designs). This book is concerned with the *analysis* of structural forms to ensure that design criteria are met.

The three major structural criteria are strength, stiffness and stability. Successive chapters are devoted to these topics. Individual members must certainly be strong enough to carry the loads they are designed to bear, but the overall strength of a complex structure may well be determined by the interaction of those members. The strength of structures is examined in Chapter 2.

Similarly, to be serviceable a structure must have displacements with acceptable limits – it must be stiff enough under the prescribed loading so that deflexions are not developed which might interfere with its design function (e.g. overhead rails in a factory building must remain sufficiently rigid to ensure that a gantry crane can operate without difficulty; an electric cable must be reachable by the pantogram of a train). Such deflexions are almost always elastic, and their calculation is explained in Chapter 3, and continued in Chapter 4.

Finally, the structure must remain stable. A familiar form of instability may be observed in the buckling of columns, but other forms are possible, and they include the instability of

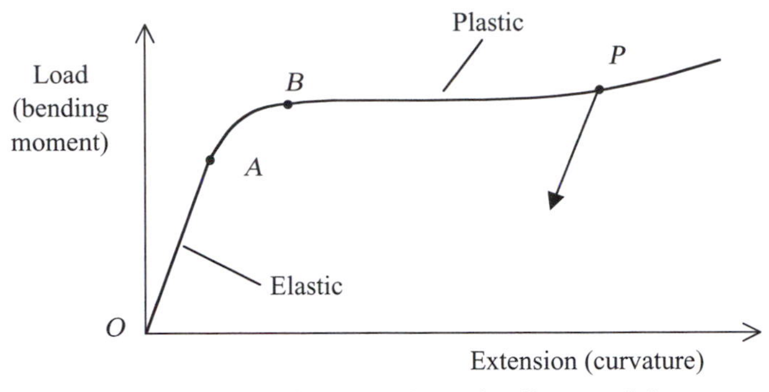

Figure 1.1. Load/extension curve for a ductile material.

the structure overall. Buckling of an individual steel or concrete member may be sudden, and could prove disastrous for the structure as a whole, although in certain types of construction (for example, plates and thin shells, which are outside the scope of this book), stable buckling can occur. Stability is studied in Chapter 5.

1.1 Structural assumptions

A first requirement of a material that is structurally useful is that it should be ductile. That is, steel, reinforced concrete (preferably under-reinforced), aluminium alloys, and perhaps wrought iron are acceptable, but cast iron and glass are not; they will shatter if incorporated as load-bearing members in a practical structure. Figure 1.1 shows schematically the results of a tensile test on a prismatic mild steel bar of a grade typically used in structural work. As the tensile load on the specimen is increased the resulting extension is at first elastic and proportional to the load (Hooke's Law), and is recoverable.

However, when the yield stress of the steel is reached the specimen extends at a more or less constant load, and behaves in a *plastic* way. If the test is interrupted at this stage, and the load is removed, the extension is not fully recoverable, and the unloading from a point such as P is elastic. The transition from elastic to plastic behaviour in such a test may be sharp, and points A and B may almost coincide. However, the important property of the schematic sketch of fig. 1.1 is that possible plastic extensions, for mild steel, are many times the extension at first yield (more than a factor of 10 before indeed the load starts to increase with the onset of strain hardening).

Such a mild steel bar is used in the example of a truss in Chapter 2, but the bar could equally be made from aluminium alloy. In that case the load/extension characteristic differs from that shown in fig. 1.1 in that portion BP of the curve would rise gently instead of being virtually horizontal. However, a design based on the load at point B of the curve would be safe for the alloy construction, and in both cases, steel and aluminium alloy, the plastic region is sufficiently large that extensions may be assumed to be unlimited, and to take place at constant load (provided there is no danger of instability; see below for the third structural assumption). The load/extension characteristic is in fact idealized as shown in fig. 1.2.

If the mild steel member is used not in tension, but in bending in the form of a beam in a structural frame, then fig. 1.1 represents – again schematically, but with some accuracy – the moment/curvature characteristic of the member. As before, the initial response is linear and elastic, but at yield large

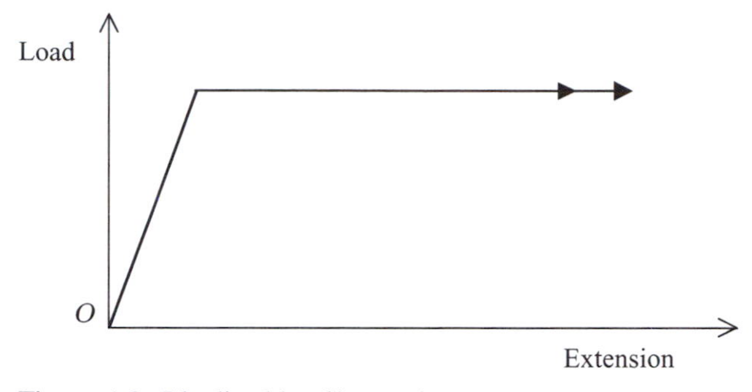

Figure 1.2. Idealised load/extension curve.

increases in curvature can occur in the beam. The yield zones are localized at *plastic hinges*; large rotations of the hinges can occur at a constant value of bending moment, defined as the *full plastic moment* of the beam. This value corresponds to the plateau *BP* in fig. 1.1. As will be seen, the formation of a single (or indeed more than one) plastic hinge does not necessarily imply that the structure has attained a limiting strength; that limit is reached when a sufficient number of hinges form so that unacceptably large deformations can occur.

A second structural assumption is concerned with the magnitude of the deformations. It is possible to construct analyses which allow for finite displacements, but the straightforward theory of structures assumes that working deformations (that is, displacements before the limiting strength is attained) are small compared with the overall dimensions of the structure. By small it is implied that changes in the overall geometry of the structure under load are negligible; thus the angles between the bars of a truss framework stay virtually unchanged, so that equilibrium equations involving these angles, and written for the undeformed structure, remain valid for the deformed state.

Finally, a third major assumption concerns the stability of structural members. This question is explored later but, essentially, care must be taken if a member is used whose load/deflexion characteristic does not exhibit the ductile plateau of the schematic fig. 1.1, but instead involves a decrease of load with increasing deformation.

1.2 Structural equations

The theory of structures is a branch of solid mechanics which deals with slightly deformable bodies, and there are only three types of basic equation which may be written to perform a structural analysis. The first set of equations expresses the static equilibrium of a structure – that is, internal structural resultants (e.g. bar forces in a truss, bending moments in a beam or frame, and so on) must be in equilibrium with the external loads acting on the structure. The familiar equations of statics – resolving forces, taking moments – are used to ensure this equilibrium. As will be seen in the next chapter, these basic equations may be used to determine the strength of a structure constructed from materials whose limiting strength (e.g. a yield stress or value of full plastic moment) is known.

The other two structural criteria – stiffness and stability – require the use of the other two sets of master structural equations. Straightforwardly, if elastic deflexions are to be calculated, then the elastic properties of the material must enter the analysis. For the trusses and beams considered in this book only the value of Young's modulus is needed. Once the value is specified for a given structural member, that member's elastic deformation can be calculated in terms of the applied internal forces (i.e. tension, compression, bending moment).

Problems involving shear deformation (which are not considered here) require the value of a second (independent) material constant, the shear modulus; this modulus is needed, for example, if the effects of twisting of a member (e.g. a steel hollow-box section) are to be investigated. (There are, in theory, 21 elastic constants for materials which possess no isotropy or other elastic symmetry. Wood, for example, has three mutually perpendicular planes of symmetry, two along the grain and one at right angles. In this case, the number of elastic constants required in theory to specify elastic behaviour is reduced to 9. However, for a reasonably homogeneous and isotropic material like steel or aluminium alloy the two constants suffice.)

Finally, the elastic deformations must be such that the members still fit together when the structure is loaded, and the structure as a whole must obey whatever boundary conditions may be specified (e.g. a beam rests on a given number of supports, a frame has its footings rigidly attached to foundations, and so on). Considerations such as these are expressed in the third set of master equations, the so-called compatibility conditions.

2 Strength

2.1 Trussed frameworks

The three equal bars shown in fig. 2.1 are supposed to be rigid and infinitely strong; they are pinned together at B and C with frictionless joints, and similar pins at A and D connect the assemblage to a rigid foundation. Evidently the figure does not represent a (two-dimensional) structure – it is a mechanism (a four-bar chain, counting the ground AD as one of the bars) incapable of carrying load. The addition of a diagonal member AC enables load to be applied – for example, the horizontal force W at joint C, fig. 2.2. The statical analysis of the truss is shown in fig. 2.3, in which the bar forces shown have been obtained by resolving horizontally and vertically at the frictionless joints. At B the two members meeting at right angles must each carry zero load, while the resolution of forces at joint C shows that the added member AC carries a tension $W\sqrt{2}$, while the (rigid) member CD is subject to a compression W. (In accordance with the assumption of small

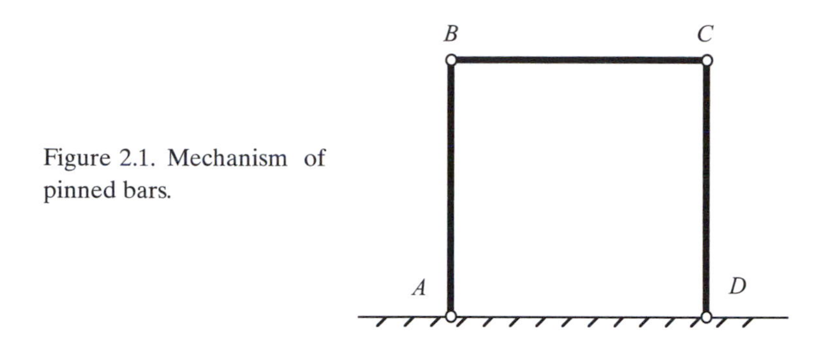

Figure 2.1. Mechanism of pinned bars.

deformations, the 90- and 45-degree angles in fig. 2.3 remain unchanged for the purpose of the resolution of forces.)

In contrast to the original three rigid members, the diagonal AC is a structural element which elongates slightly under the action of its tensile load. If the load/extension characteristic of member AC is known (that is, it has a known cross-sectional area and elastic modulus), then its extension can be calculated in terms of the force W, and the deflexion of point C may be determined.

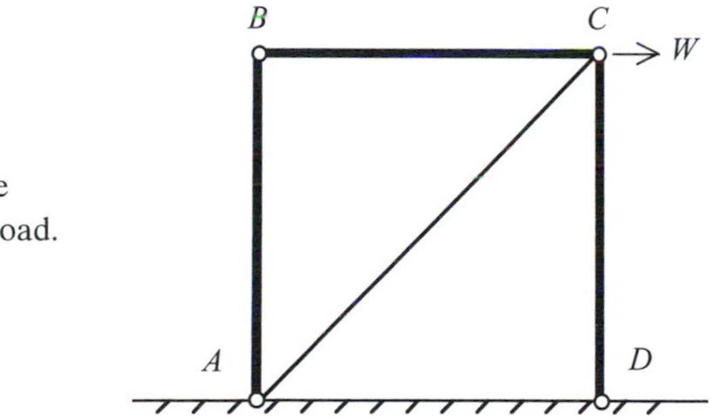

Figure 2.2. Structure capable of carrying load.

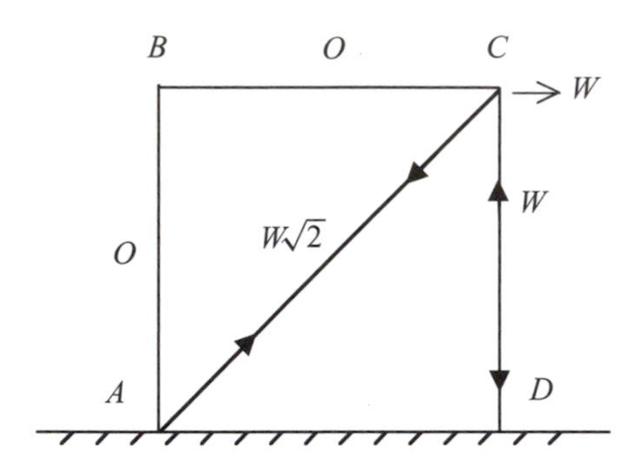

Figure 2.3. Bar forces due to load W.

The problem of the stiffness of the truss – that, is the determination of the deflexion of joint C – is discussed in Chapter 3. The present objective is to calculate the strength of the simple structure shown in fig. 2.2. If the member AC can sustain a maximum load of value T, then clearly the greatest value of W is $T/\sqrt{2}$. At this load, indefinite ductile extension of bar AC occurs in accordance with the idealized characteristic of fig. 1.2, and deflexions of the structure occur which are no longer small – a mechanism of collapse (the four-bar chain) has developed.

This analysis can hardly be dignified by the label Theory of Structures. The structural problem proper is illustrated in fig. 2.4, in which a second structural member BD has been added to the truss; as before, all joints are supposed to be freely pinned, and the two diagonals have no connexion where they cross. Under the action of the applied load W tensions P_1 and P_2 are developed in the two diagonal members, as shown. Resolution of forces at joint B leads to the marked values of tension in bars BC and BA. Tensile forces are denoted positive, so that the tension $-P_2\sqrt{2}$ marked for bar

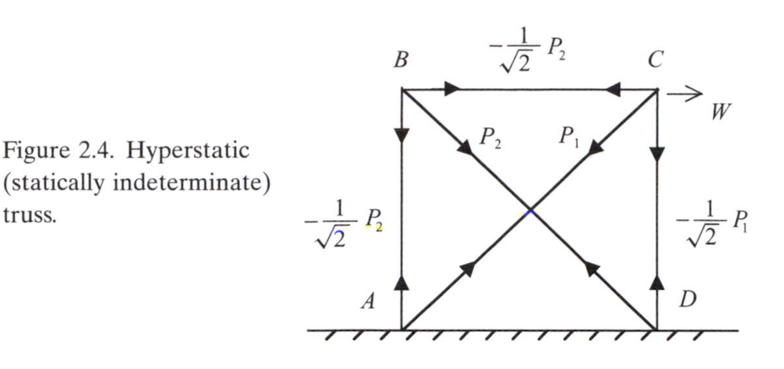

Figure 2.4. Hyperstatic (statically indeterminate) truss.

BA implies that the bar is in fact in compression. (It may be imagined that bar *BD* would be put into compression under the action of the applied load *W*, but in fact, as will be seen, it is possible for the value of P_2 to be positive.) Resolution of forces vertically at joint *C* leads to the marked value of the force in bar *CD*. Resolving forces horizontally at *C* leads to the equation

$$P_1 - P_2 = \sqrt{2}W. \tag{2.1}$$

No more information can be obtained from the use of the equations of statics, and the truss is *statically indeterminate*, or *hyperstatic*. The values of P_1 and P_2 have not been determined, although they are related by eq. (2.1). Nevertheless, it is possible to make a statement about the strength of the truss.

As before, the framing members *AB*, *BC* and *CD* are assumed to be rigid and infinitely strong. For simplicity, the two structural members *AC* and *BD* are taken to be identical, and to be ductile with maximum loads in tension *T* and in compression *C*. That is, it is assumed (at this stage) that there

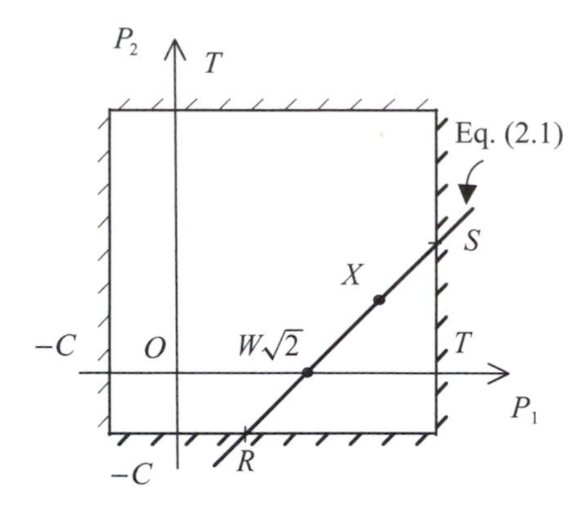

Figure 2.5. Yield surface.

is no question of instability, and that each diagonal bar can accept a maximum compressive load of value C. If a plot is made of the bar forces P_1 and P_2, fig. 2.5, then – whatever their values – any point X representing the forces must lie within the *yield surface* shown. That is, both P_1 and P_2 must lie within the range $-C$ and $+T$. Also shown in fig. 2.5 is a plot of eq. (2.1); the point X, representing the state of the truss under the load W, must lie on the line RS, which intersects the P_1 axis at the value $W\sqrt{2}$. As the value of W is increased this point of intersection moves to the right, and the maximum possible value of W is attained when the line of eq. (2.1) reaches the corner of the yield surface (see fig. 2.6). At this stage, the general point X can only just be contained within the yield surface. Comparing figs 2.5 and 2.6 shows that the value of W is given by $(C + T) = W\sqrt{2}$, and this represents the greatest load that the truss can carry. This collapse value could have been written directly from an examination of eq. (2.1); if the value of W is to be as large as possible, then P_1 must have its largest value $+T$, and P_2 its smallest, $-C$.

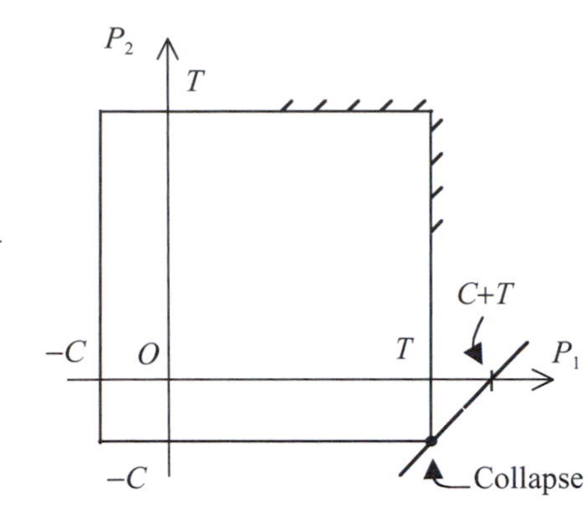

Figure 2.6. Bar forces at collapse of the truss.

Of extreme interest is the information obtainable from eq. (2.1) when there is no load on the truss; that is when $W = 0$. In this case $P_1 = P_2$, which represents a state of self-stress, characteristic of a hyperstatic structure but impossible for one which is statically indeterminate. The state of self-stress is easy to visualize for the simple truss shown in fig. 2.4. The diagonals AC and BD might be light flexible ties, for example, in one of which is installed a turnbuckle. To improve the stiffness of the truss (as will be seen in Chapter 3), the turnbuckle might be tightened to induce tension in AC, which is inescapably accompanied by the same value of tension induced in BD. Alternatively, the diagonals might be (as assumed) structural members capable of carrying compressive forces without buckling; if one of these is manufactured to be slightly too long, then it must be forced into the assembly, with resultant equal compressive forces induced in both diagonals. These two cases are illustrated in the plot shown in fig. 2.7: the point representing the state of the truss under zero external load must lie on the line $P_1 = P_2$, and could be at X for initial tension or at Y

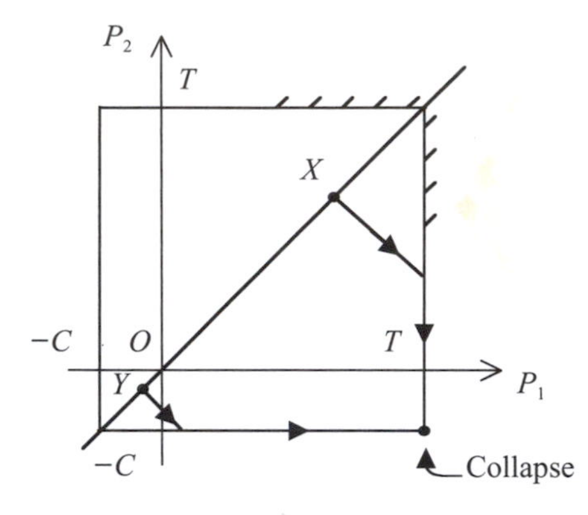

Figure 2.7. X and Y represent states of self-stress; application of load leads to the same unique collapse state.

for initial compression in the diagonals. If now the load W acting on the truss is increased slowly, then the loading point representing P_1 and P_2 moves along the line from X shown arrowed, with P_1 increasing and P_2 decreasing (the calculation of this line is given in Chapter 3). At a certain value of the load W the value of P_1 reaches T, and bar AC yields in tension. However, the load may be increased further since the force in member BD has not reached a limit – the extra load results in a progressive further decrease in the value of P_2, until it reaches the value $-C$, and the collapse corner of the yield surface is attained (as in fig. 2.6). Similarly, starting from the compressive state denoted by Y in fig. 2.7, member BD first yields in compression, and the load may be further increased until the same collapse corner of the yield surface is reached.

In general, the initial state of self-stress in a structure is not known. For this simple example, a turnbuckle may indeed be used to induce desired values of initial tension in the diagonals, but a real structure is inevitably subject to unknown manufacturing defects. If the structure is hyperstatic, the

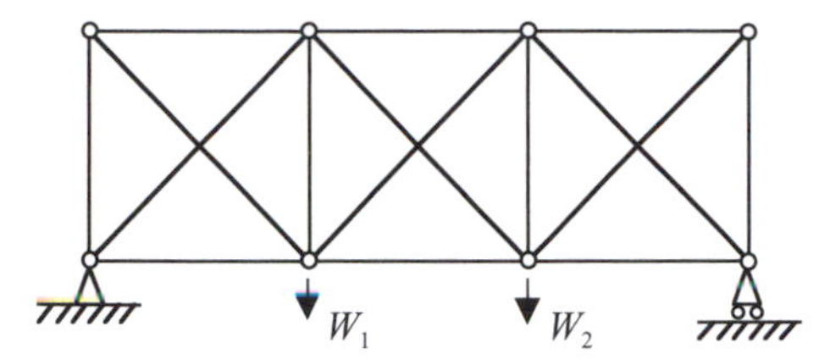

Figure 2.8. Truss with three redundant bars.

defective members must be forced into the construction, and the initial stresses in the members cannot be pre-determined. However, whatever the initial unloaded state of the structure, behaviour under loading follows the pattern illustrated schematically in fig. 2.7. Indeed, with the simple assumptions made so far, the collapse load of a ductile structure is unique, and is not affected by the unknown initial state of self-stress. Moreover, it is not necessary, in calculating the value of the collapse load, to follow the history of loading such as that illustrated in fig. 2.7; fig. 2.5 leads inexorably to fig. 2.6 to give the maximum value of W. This, of course, is an extremely powerful statement, all the more so because very small manufacturing defects can produce disproportionately large states of initial stress.

The illustrative example is deliberately simple (with just two structural members) in order to develop the ideas of analysis for strength. A more realistic, but still two-dimensional, structure is shown in fig. 2.8; this outline of a bridge truss has 16 structural members, of which 3 are redundant – that is, the truss is hyperstatic to the third degree. If, for example, one diagonal member were removed from each of the three bays of the truss, the equations of statics (i.e. resolution of forces at the joints) would enable the forces in the remaining

13 bars to be found in terms of the loads W_1 and W_2. (Note that Maxwell's Rule states that a two-dimensional truss is statically determinate if $b = 2j - 3$, where b is the number of bars and j is the number of pinned joints. Thus, in fig. 2.8, there are 8 joints, and resolution of forces at each joint, horizontally and vertically, gives 16 equations, 3 of which serve to determine the support reactions. The rule is correct, but not foolproof; the members of the truss must be arranged sensibly. If, for example, the two diagonals of the central panel of the truss in fig. 2.8 were removed, the rule would imply that the truss still had one redundancy – in fact, it is twice hyperstatic but at the same time the central panel could deform freely as a mechanism. Attention must also be given to the precise way such a truss is supported externally.)

The 16 bar forces $P_1, P_2 \ldots$ in the truss shown in fig. 2.8 can be evaluated by statics in terms of the external loads W_1 and W_2, but only in terms of three unknowns. Formally, each of the bar forces P is subject to a continued inequality of the form $-C \leq P \leq T$, where the values of C and T may well differ for each of the 16 members. The mathematical problem is to maximize – subject to the 16 continued inequalities – the value of a load factor λ, applied to each and every external load (i.e. W_1 and W_2 in this case); at this maximum value collapse occurs, as in fig. 2.6. The problem is one of linear programming – tedious if done by hand, but for which standard computing programs are available. Moreover, a computer program may be used to assist in design rather than analysis. For the truss in fig. 2.8 constructed of members of known sizes, it will be found that a regular collapse state (there are exceptions, which are illustrated later in connexion with structures in bending) involves 4 members of the truss reaching their limiting loads in either tension or compression. (Knowledge of the values of 3 bar forces would make the truss effectively statically determinate; the fourth bar gives a mechanism, by Maxwell's Rule.) The computer program can then increase the sizes of those 4 members, and the loads can be increased until eventually all 16 members could be playing their full part in contributing to the strength of the truss.

2.2 Virtual work

Equation (2.1) for the truss with one indeterminacy (see fig. 2.4), was obtained directly by considering static equilibrium of each joint in turn. The calculation was straightforward for this simple truss; as was seen, 16 such equations may be written for the more complex truss of fig. 2.8, but they are still easy to write.

However, equations such as (2.1) may be established by the use of the potent principle of virtual work. (Proof of this principle is outlined in Appendix A.) An imaginary (i.e. virtual) displacement is given to a structure which need bear no relation to the actual displacement which could occur under the given loading system. That is, a statement is made about a possible deformation pattern for the structure. Second, the statical equations are written for the structure – for the truss, the internal bar forces are deemed to be in equilibrium with the externally applied loads. Thus two of the three possible master statements of the theory of structures enter into the writing of the equation of virtual work; no use whatever is made of the third master statement, expressing the properties of the materials of which the structure is made.

In symbols, the two statements may be written:

$$\left.\begin{array}{l}\text{External loads } W \text{ are in equilibrium with}\\ \quad\text{the internal bar forces } P.\\ \text{Imaginary joint displacements } \Delta \text{ involve}\\ \quad\text{imaginary bar extensions } e.\end{array}\right\} \tag{2.2}$$

Then the principle of virtual work states that

$$\sum W \cdot \Delta = \sum Pe, \tag{2.3}$$

where W stands for $W_1, W_2 \ldots$ and so on. In this equation, the essential feature is that the two statements can be truly independent – the displacements Δ are not produced by the loads W.

Thus, for the truss shown in fig. 2.4, the equilibrium statement is expressed in the forces marked in the figure. A possible virtual displacement (indeed the only possible displacement if the outer bars AB and so on are rigid) is shown in fig. 2.9. The truss has been given a sway of magnitude Δ; evidently, this involves an extension (e_{AC}) in the distance AC which is occupied by a diagonal bar in the real truss, and similarly a compression e_{BD} in the distance BD. There are different ways of determining these imaginary bar extensions, but for the sway of fig. 2.9 a conventional displacement diagram suffices. Fixed points A and D coincide with the pole o of the displacement diagram, and are shown in lower case in fig. 2.10. The imposed

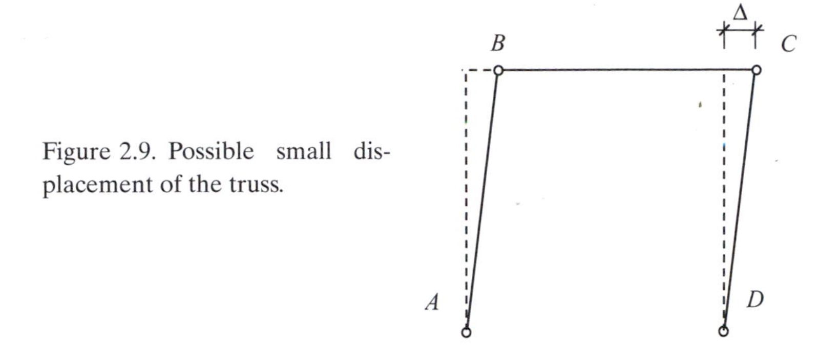

Figure 2.9. Possible small displacement of the truss.

displacement Δ locates the point *c* in the diagram, and also *b*, since bar *BD* is supposed to be inextensible. The (imaginary) extension of bar *AC* – that is, e_{AC} – occurs in the direction *oc*′ in fig. 2.10 – that is, along a line parallel to *AC* (since displacements are always assumed to be very small). However, the bar *AC* can rotate rigidly about the pin *A*, as represented by the dotted line *c*′*c* (this is actually a small arc of a circle).

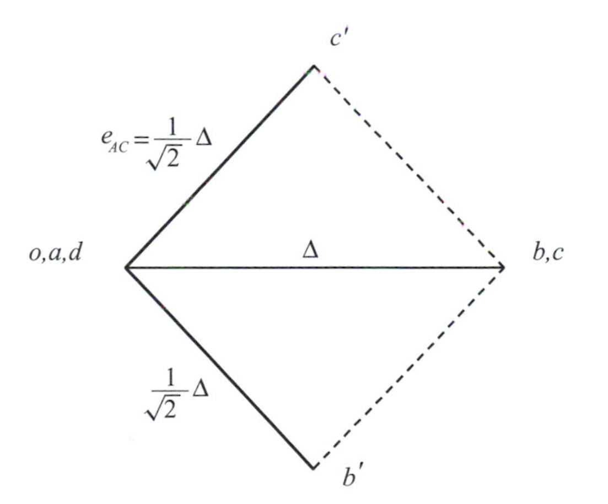

Figure 2.10. Displacement diagram corresponding to fig. 2.9.

Thus c' is located, and $e_{AC} = \Delta/\sqrt{2}$. Similarly $e_{BD} = -\Delta/\sqrt{2}$, with a negative sign because of the relative positions of o and b' in fig. 2.10. The statements (2.2) can then be written as follows:

$$\left.\begin{array}{l}\text{The load } W \text{ is in equilibrium with bar}\\ \quad\text{forces } P_1 \text{ and } P_2.\\ \text{Displacement } \Delta \text{ is compatible with bar}\\ \quad\text{extensions } \Delta/\sqrt{2} \text{ and } -\Delta\sqrt{2},\end{array}\right\}$$

so that eq. (2.3) gives

$$W\Delta = P_1\Delta/\sqrt{2} - P_2\Delta/\sqrt{2}, \qquad (2.4)$$

which, of course, is precisely eq. (2.1).

The analysis of displacements of trusses is not trivial, and displacement diagrams for more elaborate trusses (such as that shown in fig. 2.8) can become very complex. As will be seen in Chapter 3, truss analysis is simpler if the equation of virtual work is used the other way around. That is, instead of real equilibrium systems being connected with imaginary displacements, the terms in eq. (2.3) are interpreted as referring to the actual displacements and to artificial systems of equilibrium. For trusses, difficult problems of geometry are then replaced by easier problems of statics.

2.3 Structures in bending

The entire discussion of this chapter may be rewritten in terms of beams and frames (i.e. structures subjected to bending)

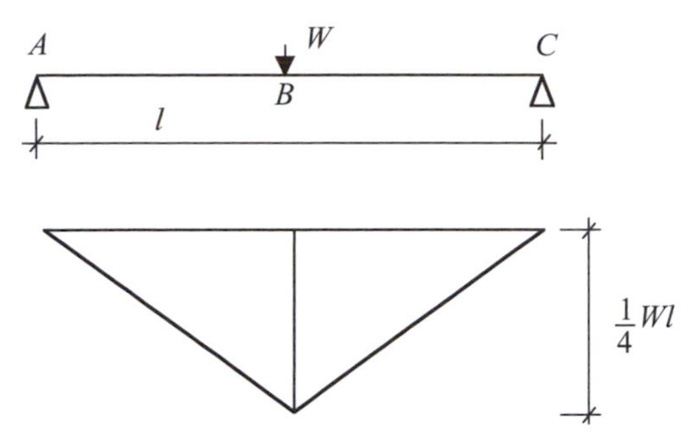

Figure 2.11. Simply supported beam.

rather than in terms of trusses (whose members act in tension and compression). The simply supported beam shown in fig. 2.11 is statically determinate, and the bending-moment diagram is as shown, with a maximum value of moment $Wl/4$ at the midpoint B of the beam. If the beam is of uniform section having full plastic moment M_P, then the largest value of W is given by $Wl/4 = M_P$.

In fig. 2.12 the same beam is shown, but it is now clamped at the left-hand end A. The clamp is normally assumed to impose zero slope on the beam at A, but in fact, as far as strength is concerned, all that is necessary is that the clamp be sufficiently strong and stiff so that the full plastic moment of the beam can be developed. Under the load W a bending moment M_A acts at A and M_B at B, both shown hogging in fig. 2.12(b); it is to be expected that M_B has a negative value.

The complete bending-moment diagram for the beam may be determined conveniently by combining the free bending moments for the equivalent simply supported beam, fig. 2.12(c), with the reactant moments due to the clamping

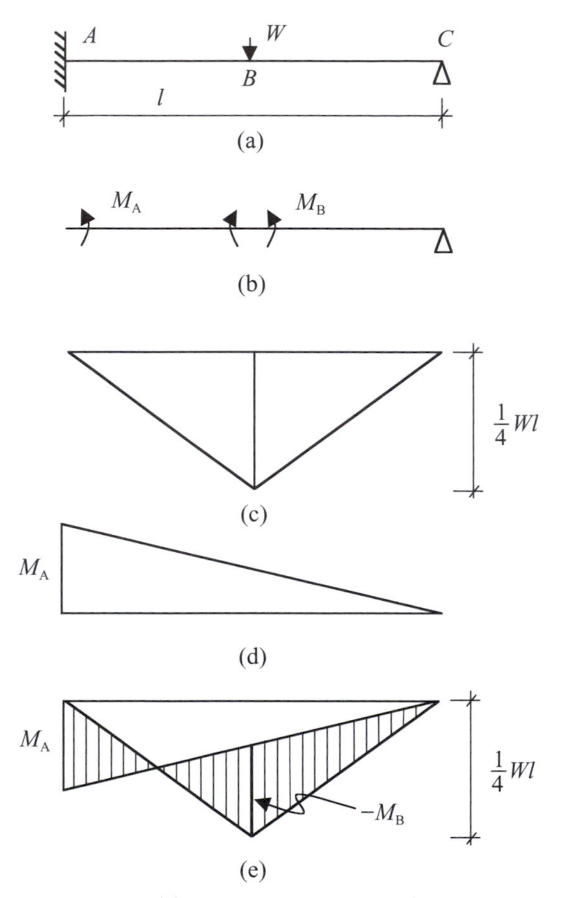

Figure 2.12. (a) Propped cantilever, (b) bending moments induced by load, (c) free bending-moment diagram, (d) reactant bending-moment diagram, (e) general bending-moment diagram.

moment M_A, fig. 2.12(d). This superposition may be done analytically (as it would be in a computer program), but it is shown graphically for this simple problem in fig. 2.12(e). Evidently,

$$\frac{1}{2}M_A - M_B = \frac{1}{4}Wl, \tag{2.5}$$

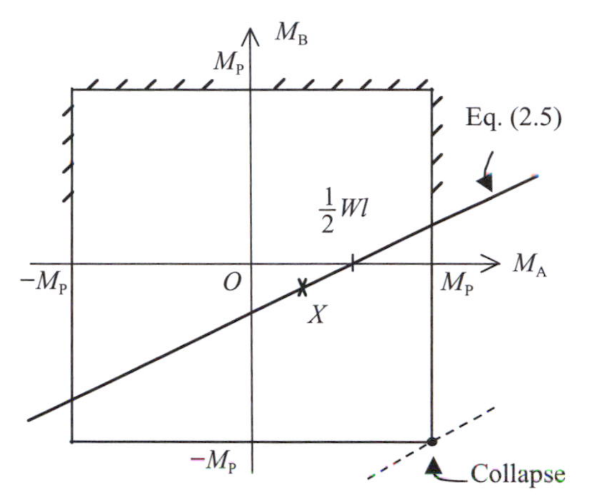

Figure 2.13. Yield surface for beam in fig. 2.12.

and this is the only information that can be obtained by considering the statics of the problem (cf eq. (2.1) for the truss). However, the values of M_A and M_B are constrained to lie within the yield surface shown in fig. 2.13; those values, represented by the point X, must lie on the plot of eq. (2.5) shown in the figure. The intersection of the line with the M_A-axis fixes its position for a given value of W; if that value is to be made as large as possible, then the state of the beam is given by $M_A = M_P$, $M_B = -M_P$. Once again, this result can be seen directly from an examination of eq. (2.5), if the value of W is to be made as large as possible. The maximum value is

$$W = \frac{6M_P}{l}. \tag{2.6}$$

In this state, the beam is on the point of collapse by the formation of the two plastic hinges in line with the real hinge

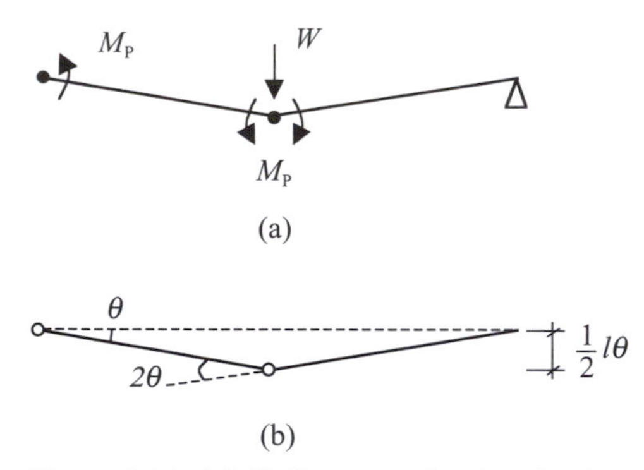

Figure 2.14. (a) Collapse mechanism for beam in fig. 2.12, (b) virtual mechanism.

at the prop at the right-hand end, fig. 2.14(a), this being an elementary mechanism capable of (infinitesimal) displacement.

The collapse eq. (2.6) can be written directly from fig. 2.14(b) by equating the work done by the external load W to the work dissipated in the plastic hinges:

$$M_P(\theta) + (-M_P)(-2\theta) = W\left(\frac{1}{2}l\theta\right), \tag{2.7}$$

which leads directly to eq. (2.6).

Care has been taken with signs in writing eq. (2.7), hogging movements and rotations being denoted as positive, and sagging as negative. In fact, work dissipated in plastic-hinge rotations is always positive – a negative value of moment is accompanied by a negative hinge rotation. However, to describe the general rather than the collapse state of the beam, fig. 2.14(b) may be used as a *virtual* mechanism in conjunction with the equilibrium state indicated in fig. 2.12(b). As shown

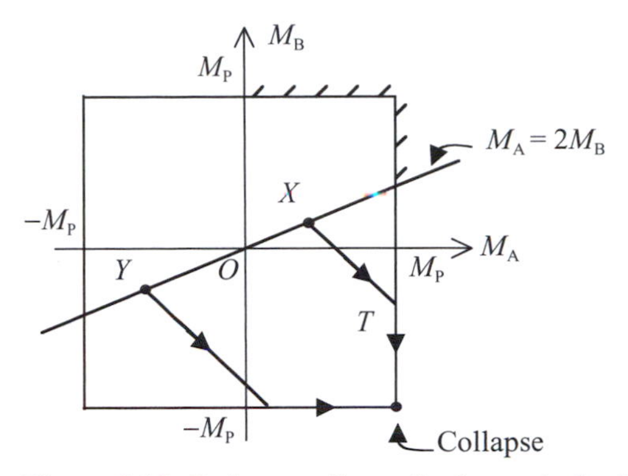

Figure 2.15. Paths to collapse for beam in fig. 2.12, starting from states of self-stress.

in Appendix A, the virtual-work equation for structures in bending becomes [cf eq. (2.3)]:

$$\sum W \cdot \Delta = \sum M\theta, \qquad (2.8)$$

so that

$$W\left(\frac{l}{2}\theta\right) = (M_A)(\theta) + M_B(-2\theta), \qquad (2.9)$$

which is the basic equilibrium eq. (2.5).

The beam is hyperstatic; before any load is applied, there could be a state of self-stress given by $M_A = 2M_B$, from eq. (2.5) with $W = 0$. Such a state would arise, for example, if the supports at ends A and C of the beam were not at the same level; in fig. 2.15, the initial state would be represented by point X if the prop at the right-hand end were too low, and by Y if it were too high. If loading of the beam starts from point X, then the arrowed path would be followed (again, the calculation of this path is given in Chapter 3); a hinge would form first at the clamped end A ($M_A = M_P$) and further increase of load would be accompanied by rotation of that hinge until the value of M_B reaches $-M_P$. Similarly, loading from point Y implies that a hinge forms first at B, under the load, with the moment at A finally reaching the full plastic value. As for the truss, the final collapse state is independent of the (unknown) state of initial self-stress.

2.4 Plastic theory

The calculation of the strength of structures is, of course, the objective of what is now known as plastic theory. The theory is usually formulated in terms of bending structures – beams and frames – although early presentations were indeed for trussed frameworks (with warnings about the instability of compression members). In what follows, the bending problem is discussed, but the conclusions – above all, the bound theorems – apply equally to any ductile structure satisfying the assumptions outlined in Chapter 1.

In practice, a structure is acted on by a number of specified loads. For a conventional elastic analysis, each load may be considered separately, and the elastic response, and consequent elastic stresses, calculated at each critical section of the structure. These stresses can then be summed to give greatest and least values, and a suitable safety factor incorporated to guard against yield at the critical sections. For example, the

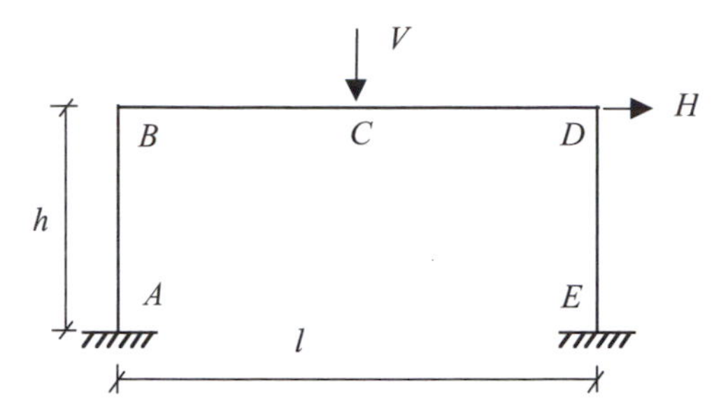

Figure 2.16. Simple rectangular portal frame.

idealised fixed-base portal frame shown in fig. 2.16 is acted on by two loads, where V might represent self-weight and a possible snow load, and H the action of wind blowing from left to right. The frame is assumed to have full-strength connexions at B and D, and to the footings at A and E. The greatest (hogging) moment at critical section D of the frame occurs when V and H act together, whereas the greatest moment at section B occurs under the action of V alone.

For a plastic analysis, the designer (or the computer program) arranges the loading in a way which leads to the most critical condition for the structure as a whole. In fig. 2.16, for example, if the worst combination involves the action of both loads, then the portal frame would be analysed accordingly. Instead of a safety factor based on stress, the concept of a load factor is introduced. Both loads V and H are imagined to be increased by the *same* factor λ; for a frame built of members of known strength, the largest value of λ is the collapse load factor (for straightforward simple steel structures, a collapse load factor of value 1.75 is appropriate).

In fact, it is possible to carry out plastic analysis with a range of loads – each load subject to a different load factor – and also to analyse the response of structures to random and repeated application of such loads, each of which might be specified to lie within certain limits (e.g. wind blowing from left to right, not at all, or from right to left). For simplicity, the following presentation is limited to the case in which all loads are subject to the same value of load factor. However, a statistical element may be introduced. The safety factors of stress used in elastic design are allowed different values to reflect the fact that it is unlikely, for example, that a crane is operating in a factory building at the same time that a high wind is blowing and that there is a full snow load on the roof. Such statistical ideas find their way into plastic design. In fig. 2.16, the analysis under the vertical load V alone could be carried out at a load factor of 1.75, while the analysis under both loads might be subject to a lower factor of 1.4; naturally, the designer is constrained by the more critical of these two cases.

Plastic analysis uses, perforce, the three master statements of structural theory, which are formulated as the requirements of equilibrium, yield and mechanism. The equilibrium condition is no more than the familiar equations of statics – for beams and frames, the internal bending moments and external loads must together satisfy these equilibrium equations. Material properties enter the analysis by the simple statement that the value of bending moment at any section must not exceed the value of the full plastic moment at that section – this is the yield condition. Where the value of the full plastic moment is reached a plastic hinge is formed, capable of rotation under the constant value of the full plastic moment. The mechanism condition is the vestigial representation of deformation; there must be some arrangement of (ductile) plastic hinges to permit a mechanism of one degree of freedom (under certain circumstances, possibly more than one degree).

The basic plastic theorems derive from these three requirements of equilibrium, yield and mechanism, and their proofs are outlined in Appendix B. The theorems may be displayed as:

$$\lambda = \lambda_c \begin{cases} \left.\begin{array}{l}\text{Equilibrium}\\ \text{Yield}\end{array}\right\} & -\text{ Safe theorem,} \quad \lambda \leq \lambda_c \\ \text{Mechanism} & -\text{ Unsafe theorem,} \quad \lambda \geq \lambda_c \end{cases} \tag{2.10}$$

In words, the uniqueness theorem states that if all three conditions are satisfied for a given structure at a load factor λ, then the value of the load factor is unique, and λ is equal to the collapse load factor λ_c. (The corresponding collapse mechanism is not necessarily unique; under certain circumstances, alternative patterns of hinges may lead to the same value of collapse load factor.) By contrast, if a mechanism is studied without reference to the other conditions, then the value of load factor (perhaps calculated, as will be seen, from a work equation) is an over-estimate of the true collapse load factor, and is, from the point of view of design, unsafe. Finally, if only the conditions of equilibrium and yield are considered, then

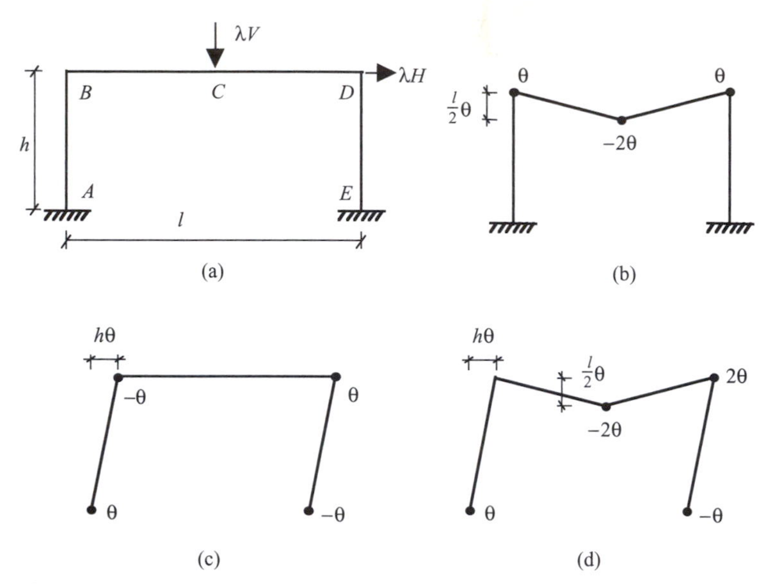

Figure 2.17. Possible collapse mechanisms for portal frame.

a safe estimate may be made of the strength of the structure. These theorems may be illustrated by examination of the rectangular portal frame shown in fig. 2.16.

The frame is redrawn in fig. 2.17(a), together with sketches of three possible collapse mechanisms. The bending-moment diagram for the portal frame under the given loading consists of straight lines between the cardinal points $A \ldots E$; these are the sections at which plastic hinges might form. In general, knowledge of the values of the bending moments $M_{\mathrm{A}} \ldots M_{\mathrm{E}}$ enables the complete bending-moment diagram to be constructed.

If the mechanisms of fig. 2.17 are interpreted, in the first instance, as virtual rather than collapse mechanisms, then the virtual-work eq. (2.8) may be used to provide relationships between the values of the bending moments at the five cardinal points. The beam mechanism of fig. 2.17(b) leads to

$$M_B - 2M_C + M_D \qquad = \lambda \frac{Vl}{2} \tag{2.11}$$

(where the parameter θ has been cancelled throughout), while the sway mechanism of fig. 2.17(c) gives

$$M_A - M_B \qquad + M_D - M_E = \lambda H h. \tag{2.12}$$

These equilibrium equations must always be satisfied, whether the frame is in an elastic or a plastic state; three values of bending moments (e.g. M_A, M_B and M_C) may be assigned, but the other two are then calculable from eqs (2.11) and (2.12).

The three bending moments M_A, M_B and M_C could be considered as the three redundancies of this hyperstatic frame, and it would be the objective of an elastic analysis to calculate their values. The plastic-collapse equations, however, can be written at once. If fig. 2.17(b) is now interpreted as a plastic-collapse mechanism and if the frame is of uniform section having full plastic moment M_P, then the plastic work equation (for a small displacement θ in the collapse state) gives

$$\lambda \frac{Vl}{2}\theta = (M_P)(\theta) + (-M_P)(-2\theta) + (M_P)(\theta)$$

$$\text{or } \lambda \frac{Vl}{2} = 4M_P, \tag{2.13}$$

while fig. 2.17(c) leads to

$$\lambda H h = 4M_P. \tag{2.14}$$

Equations (2.13) and (2.14) can, of course, be derived from eqs (2.11) and (2.12) by maximizing the value of the load factor λ, setting the bending moments $M_A \ldots M_E$ equal to $\pm M_P$, as appropriate.

Since the frame has 3 redundancies and there are 5 values of bending moment to be determined, eqs (2.11) and (2.12) are the *only* independent equilibrium equations that can be written. Figure 2.17(d) does indeed show a different mechanism, and the equation of virtual work leads to

$$M_A \qquad - 2M_C + 2M_D - M_E = \lambda\left(Hh + \frac{Vl}{2}\right). \qquad (2.15)$$

It is clear, however, that eq. (2.16) is not an independent equilibrium equation, but merely the summation of eqs (2.11) and (2.12); indeed, fig. 2.17(d) is the pictorial summation of figs 2.17(b) and (c). The plastic collapse equation corresponding to eq. (2.15) is

$$\lambda\left(Hh + \frac{Vl}{2}\right) = 6M_P. \qquad (2.16)$$

There are no other arrangements of hinges which give rise to mechanisms; thus there are three (and only three) estimates of the value of λ, given by eqs (2.13), (2.14) and (2.16). The unsafe theorem states that the correct value of the load factor, λ_c, is the smallest of the three estimates.

The three collapse equations are plotted in fig. 2.18 to form a yield surface, this time with loads as axes rather than internal stress resultants. The point X in the diagram represents the design (working) values of the loads V and H; it is a consequence of the plastic theorems that if point X lies within the boundaries of the yield surface, then the corresponding

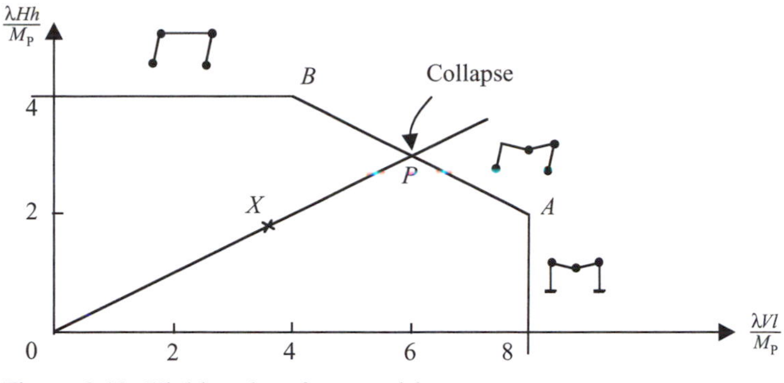

Figure 2.18. Yield surface for portal frame.

loads V and H are a safe combination – the portal frame will not collapse. Point X corresponds to a load factor of unity; as λ is increased, X moves outwards along the straight line OX, until the boundary is reached at P. The ratio OP/OX is the collapse load factor; in the case shown in fig. 2.18, collapse occurs by the formation of the mechanism in 2.17(d).

Figure 2.18 illustrates in fact one quarter of a (symmetrical) yield surface; the full boundary is obtained if negative values of V and H are considered. Such yield boundaries (in n-dimensional space if there are n loads) are closed surfaces around the origin, and are convex (no re-entrant angles). The property of safety (X lies within the yield surface) is of course of crucial importance in structural design; there has been no mention in the above discussion of any state of possible initial pre-stress. The mechanism of collapse and the value of the collapse load (or of the collapse load factor) are not affected at all by pre-stress, exactly as was seen for the truss in fig. 2.4 (collapse in fig. 2.7) or for the beam in fig. 2.12 (collapse in fig. 2.15).

The determination of the boundary of a yield surface by consideration of unsafe mechanisms of collapse gives a

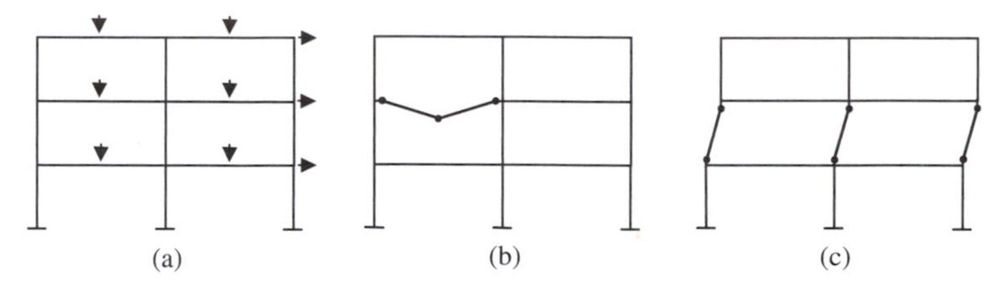

Figure 2.19. Multi-storey, multi-bay frame with 18 redundancies.

powerful algorithm for calculation (historically by hand, but now by computer). If a framed structure has a number R of redundancies, and a number N of critical sections at which plastic hinges might form, then there must exist $(N - R)$ equilibrium equations, such as eqs (2.11) and (2.12), which connect the magnitudes of the bending moments at the critical sections with the values of the external loads. Thus the frame shown in fig. 2.19 has 36 critical sections, 2 at the ends and one in the centre of each of the 6 beams, and one at each end of the 9 column lengths. The frame has 18 redundancies, and there must exist, therefore, 18 independent equilibrium equations. Now it was seen that an equilibrium equation could be derived from a mechanism; therefore there must be 18 independent mechanisms for the frame. Six are of the elementary beam type, one of which is shown in fig. 2.19(b), and there are 3 sways, of which one is shown in fig. 2.19(c). The remaining 9 relationships are degenerate joint mechanisms, expressing the fact that the moments acting on the ends of members meeting at a joint must sum to zero. From these elements can be built up a highly complex collapse mechanism.

The hinge arrangement of fig. 2.19(b) represents a partial collapse mechanism; for complex frames, it is highly likely that collapse occurs with a large part of the frame remaining statically indeterminate. The frame in fig. 2.19(b) retains 16 of the original 18 redundancies, and there is no way that the actual values of these redundancies can be found, since they are determined by movements of the external environment which are unpredictable and over which the designer has no control. A computer program will, however, provide *possible* values for the redundancies and, if the collapse solution is to be valid, these values must satisfy the yield condition – that is, the values of bending moment at those critical sections not involved in the collapse mechanism cannot exceed the values of full plastic moment at those sections.

Analysis of the simple rectangular portal frame illustrates the problem. The partial collapse mechanism in fig. 2.17(b) leaves the frame with one redundancy. Collapse occurs according to eq. (2.11) with $M_B = M_P$, $M_C = -M_P$ and $M_D = M_P$, and, as was seen, the value of the collapse load V is given by eq. (2.13). Equation (2.12) must always be satisfied, in which now $M_B = M_D = M_P$ and, so that

$$\lambda Hh = M_A - M_E. \tag{2.17}$$

Unique values of M_A and M_E cannot be found, but they must satisfy the condition that, numerically, neither can exceed M_P, so that

$$\lambda Hh \leq 2M_P. \tag{2.18}$$

Thus the collapse mode of fig. 2.17(b) can occur if

$$\lambda Hh \leq 2M_P = \lambda \frac{Vl}{4}, \tag{2.19}$$

using the collapse eq. (2.13); that is

$$Hh \leq \frac{Vl}{4}. \tag{2.20}$$

The equality sign in eq. (2.20) is represented by the corner A of the yield surface in fig. 2.18; should H exceed this value, then the collapse mode switches to that in fig. 2.17(d). The second corner B of the yield surface, $Hh = Vl$, represents the transition to the mode in fig. 2.17(c).

It was noted that the mechanism method outlined above can be programmed for the computer. Equally, the safe approach, involving equilibrium and yield, is merely (as mentioned earlier) an exercise in linear programming. For the rectangular portal frame the two equilibrium eqs (2.11) and (2.12) must hold, and to these are added the requirements

$$-(M_{\mathrm{P}})_i \leq M_i \leq (M_{\mathrm{P}})_i \tag{2.21}$$

where M_i stands in turn for $M_{\mathrm{A}} \ldots M_{\mathrm{E}}$, and where, in general, the value of M_{P} need not be the same at each critical section. The value of λ in eqs (2.11) and (2.12) must be maximized subject to the restraining inequalities (2.21).

The determination of the final value of λ results from the *analysis* of a given frame under specified loading. For *design*, the value of λ is specified; what is required is the value of M_{P} for a frame which just achieves the required strength (for a non-uniform frame, the values of M_{P} vary from member to member). The problem is inverted; for example, the unsafe theorem states that if all possible mechanisms of collapse are studied, the one that is correct gives the *greatest* design value(s) for M_{P}.

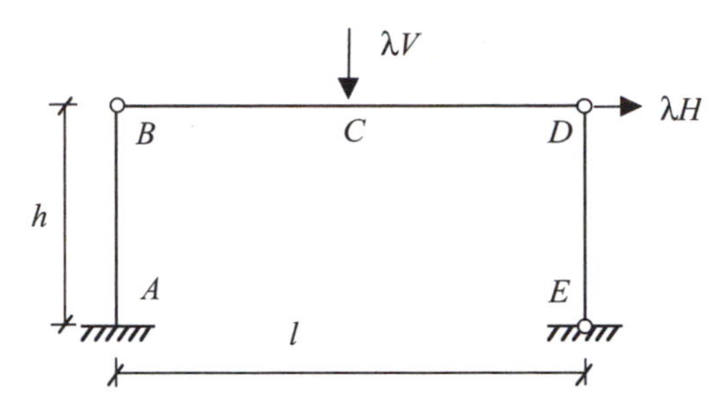

Figure 2.20. Statically determinate frame.

Indeed, for design purposes, the value of the proportional load factor need not be introduced at all until the very last stage. The frame is designed to collapse under the specified working values of the loads – that is, at unit load factor. The corresponding values of M_P are then determined. All these values of M_P are then increased by the same chosen load factor; if the factor were 1.75, then the members would have strengths 1.75 times the values resulting from the design exercise.

As a final aid to calculation, external loads may be introduced as an internal set of bending moments in equilibrium with those loads (in the way that fixed-end moments may be used in elastic analysis, as will be seen in Chapter 4). Any such set will suffice, and a simple way to construct a set is to consider an equivalent statically determinate frame. The rectangular portal frame has been modified by the insertion of three frictionless pins (see fig. 2.20) and a set of equilibrium moments can be written:

$$M_i^{\circ} \equiv (M_A^{\circ}, M_B^{\circ}, M_C^{\circ}, M_D^{\circ}, M_E^{\circ}) \equiv \left(\lambda Hh, 0, -\lambda\frac{Vl}{4}, 0, 0\right). \tag{2.22}$$

Now the bending moments in the original frame can be written:

$$M_i = M_i^\circ + m_i, \tag{2.23}$$

where m_i are (unknown) self-stressing moments, in equilibrium therefore with zero external load. [From eq. (2.11), for example, $m_B - 2m_C - m_D = 0$.] Equation (2.23) is illustrated graphically in fig. 2.12 for the propped cantilever, where it was described as the superposition of free and reactant moments.

If a mechanism of collapse θ_i is examined, then it is known that, for this postulated mechanism, the bending moments at the hinge locations have value M_P, so that $M_i = |M_P|$ at these hinges. On multiplying through by the values of the hinge rotations, and summing for the whole mechanism,

$$\sum M_i\theta_i = \sum (M_i^\circ + m_i)\theta_i = \sum (M_P)_i\,|\theta_i|. \tag{2.24}$$

As noted above, $\sum m_i\theta_i = 0$, since the self-stressing moments are in equilibrium with zero external load, so eq. (2.24) becomes

$$\sum M^\circ\theta = \sum M_P\,|\theta|, \tag{2.25}$$

where the suffix *i* has been dropped. For example, the set of bending moments given in (2.22) used with the hinge rotations in fig. 2.17(b) leads directly to eq. (2.13).

Finally, should a computer program furnish an elastic distribution $\mathcal{M}$ of bending moments (which must of necessity be a special case of an equilibrium distribution), then eq. (2.25) leads to

$$\sum \mathcal{M}\,\theta = \sum M_{\mathrm{P}}\,|\theta|; \tag{2.26}$$

a plastic collapse analysis may be made directly from an elastic solution. The distribution $\mathcal{M}$ cannot be observed by measurements on a real hyperstatic structure; as noted throughout this chapter, the actual bending moments result from the superposition of the actions of the external loads on an initial, unknown and unknowable, state of self-stress. However, whatever the starting state, the same collapse corner of the yield surface is always reached and this is effectively demonstrated by eq. (2.26).

2.5 Masonry

Stone (like brick) is essentially a brittle material; it has adequate compressive strength, but apparently lacks the ductility to serve as a useful structural material. However, as will be seen, the assembly of stone into a masonry structure, such as an arch, creates a form which has its own peculiar property of plastic deformability. Indeed, the structural theory of masonry can be embraced within plastic theory.

Figure 2.21(a) shows a schematic but, in fact, a reasonably realistic masonry arch formed from wedge-shaped stone voussoirs. The mortar between the voussoirs is usually very weak in tension, and in practice may be absent, so that, in reality, there is nothing to prevent the stones from pulling apart.

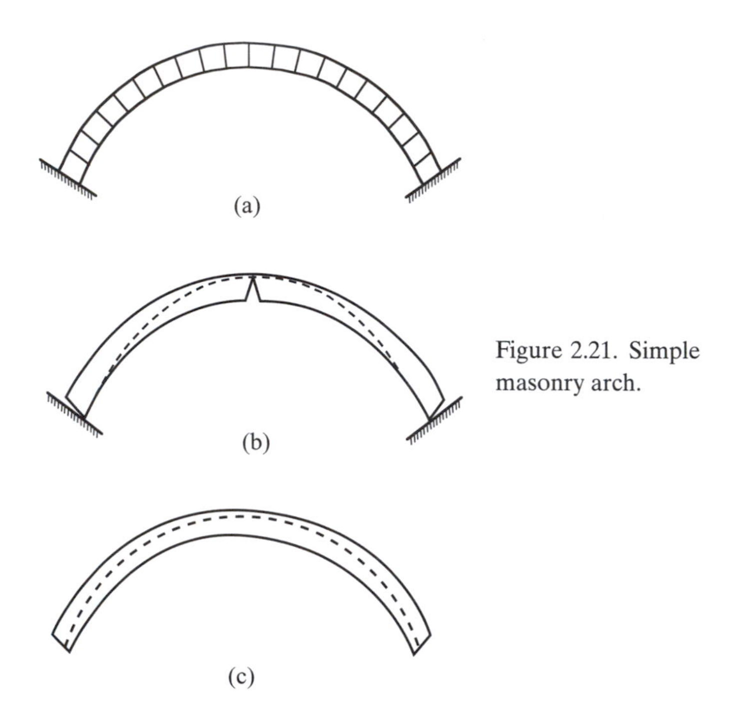

Figure 2.21. Simple masonry arch.

By contrast, the compressive forces between the voussoirs are so small that the resulting stresses are very low – even a large-span bridge experiences stresses well under 10 per cent of the crushing strength of the stone. With these remarks in mind, the material properties of masonry may be formulated. The following three key assumptions are stated for the voussoir arch, but are applicable to other structural forms (e.g. the elements – towers, spires, vaults, buttresses and so on – of a great church).

Sliding failure cannot occur

It is assumed that friction is high enough between voussoirs, or that the stones are otherwise effectively interlocked, so that they cannot slide one on another. This is a reasonable assumption, although it is certainly possible to find occasional evidence of slippage in a masonry structure.

Masonry has no tensile strength

Stone itself has a definite tensile strength, but it is the joints between the stones that are weak. Thus the assumption implies that only compressive forces can be transmitted between masonry elements. In accordance with common sense, and with the principles of the plastic theorems (discussed later), this assumption is safe.

Masonry has an infinite compressive strength

This assumption is a consequence of the fact that, in practice, stresses are far removed from the crushing strength of the material. The assumption is obviously unsafe, but it is not unrealistic; it is discussed further later.

Thus a picture emerges of masonry as an assemblage of stones shaped to pack together into a coherent structural form, with that form maintained by compressive forces transmitted within the mass of the material. (Since stresses are low, the term *masonry* includes not only stone and weaker bricks, but also, say, breeze blocks and more primitive materials, such as sun-dried mud.) The question then arises as to how such a masonry assemblage might fail in any meaningful structural sense. If the masonry is infinitely strong, then it would seem that a calculation of levels of compressive stress is not relevant. However, the idea that tension is not permissible is significant.

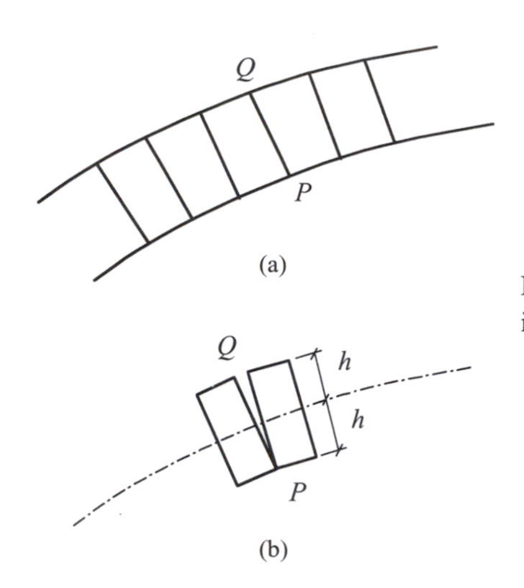

Figure 2.22. Hinge formation in masonry arch.

The arch shown in fig. 2.21(a) would be constructed on temporary falsework or centering. When the keystone has been placed, the centering may be removed, and the arch immediately starts to thrust against its abutments; those abutments (the river banks) inevitably give way. The arch – composed of strong, virtually rigid voussoirs – must adapt to a slightly increased span, and it does this by cracking at the joints, shown greatly exaggerated as hinges in fig. 2.21(b). Thus the arch is freely deformable to conform to the new span, and, despite the brittleness of the individual elements, the structure as a whole exhibits ductile behaviour.

The compressive structural forces must of necessity pass through the hinge points of fig. 2.21(b), and the broken line in the figure represents what may loosely be called the line of thrust – that is, the resultant of the compressive forces passed from voussoir to voussoir within the masonry profile. If a particular joint PQ is examined (fig. 2.22), then the structural

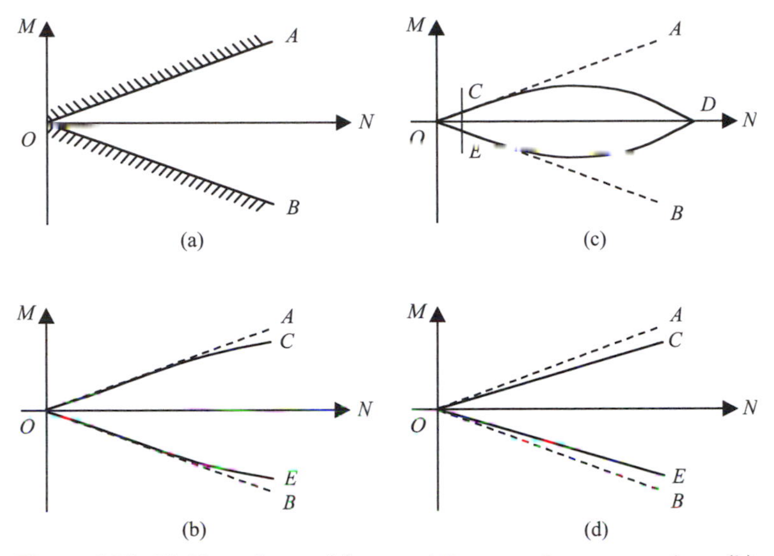

Figure 2.23. Yield surfaces: (a) material strong in compression, (b) allowing for crushing of material, (c) enlarged portion of yield surface, (d) approximation to yield surface.

action at the joint is specified in terms of the magnitude, direction and point of application of the force transmitted across the joint. The tangential component of the force is not of importance, since slip is assumed not to occur – what is needed is the value N of the normal force across the joint together with the value of its eccentricity e from the centre line. It is convenient to work with a bending moment $M = Ne$ as a second variable, so that the stress resultants M, N define the state of the arch at any particular section.

The hinge shown in fig. 2.22(b) forms when the eccentricity e of the normal thrust just has the value h; that is, when $M = hN$. The lines $M = \pm hN$ are shown as OA and OB in fig. 2.23(a), and they represent, for any given joint between voussoirs, the condition that a hinge is in existence at that joint. A general point (*N*, *M*) in the figure which lies within the open triangle *AOB* represents a thrust between voussoirs at an eccentricity less than *h*; that is, the line of thrust lies within the voussoirs at that joint and no hinge is forming. If the general point lies on *OA* or *OB*, then a hinge is forming in either the intrados or extrados of the arch. The general point cannot lie outside the region *AOB*, since this would imply tension at the joints.

The construction shown in fig. 2.23(a) involves the assumption that the material has infinite compressive stress. As the line of thrust approaches the edge of a voussoir so the stress on the diminishing area of contact increases, and a real stone with a finite crushing strength does not permit the line of contact at a hinge that is illustrated in fig. 2.22(b). Thus the lines *OA* and *OB* in fig. 2.23(a) cannot quite be reached; they are replaced by the slightly curved lines shown in fig. 2.23(b). The full boundary is formed by the parabolic arcs *OCD* and *OED* in fig. 2.23(c), and the general point (*N*, *M*) must lie within this boundary. The assumption of low mean compressive stress, in fact, constrains the point (*N*, *M*) to lie within an area such as *OCE* in fig. 2.23(c), and it is this area which is enlarged in fig. 2.23(b).

The sketches shown in fig. 2.23 are again yield surfaces of plastic theory, and plastic principles may be applied. A general point (*N*, *M*) lying within the full yield surface of fig. 2.23(c) represents a safe state for the masonry. The curved boundaries of fig. 2.23(b) also represent a safe yield surface, and they may be straightened by the device shown in fig. 2.23(d). If the mean compressive stresses are known not to exceed 10 per cent of the crushing strength of the stone, then the straight lines OA and OB may be replaced by $M = \pm 0.9hN$. Thus the real arch having a (local) ring depth of $2h$ is replaced, for the purposes of analysis, by a hypothetical arch of depth $2(0.9h)$. This kind of shrinking is important in assessing the safety of masonry arches.

The abutments of the arch shown in fig. 2.21 were imagined to move apart, leading to the pattern of cracks illustrated in fig. 2.21(b). The originally hyperstatic structure (with three redundancies) has been transformed into a three-pin arch, which is now statically determinate; for the known loading, the value of the abutment thrust may be determined (and is, in fact, the lowest value which maintains the arch in equilibrium). The three-pin arch is a well-known satisfactory structural form – the development of three hinges by cracking of the joints does not presage collapse.

If the abutments of the arch do not move apart, or move apart only slightly, and the voussoirs are almost but not absolutely rigid, then the joints between voussoirs remain tight, and no hinges occur to help locate the line of thrust. All the designer may be able to show is that the line of thrust occupies some such position as that shown in fig. 2.21(c), where it lies completely within the masonry. In fact, this is all the designer needs to show. If any one position such as that shown in fig. 2.21(c) can be found, then this is absolute proof – by the safe theorem of plastic theory – that the structure is safe. If the designer can determine a way in which the structure can carry the given loads, then the structure can certainly also find a way.

This anthropomorphic statement does not, of itself, give any indication of how safe the structure might be. Since the masonry has been assumed to be of infinite compressive strength, there is no question of failure of the material. Instead, a geometrical criterion can be devised. As shown in fig. 2.21(c), the shape of the line of thrust is not the same as the shape of the profile of the arch, and there is a minimum thickness of the arch which only just contains the line of thrust.

In 1675 Robert Hooke identified the shape of the line of thrust by his statement: 'as hangs the flexible line, so but inverted will stand the rigid arch'. In other words, if the given loading for the arch were applied to a light string, then the shape of that string, in tension, would be the same, inverted, as that of the arch to carry the same loads in compression. In fig. 2.21(c), for example, if the loading resulting from the circular profile of the arch were uniform, then the line of thrust shown in the figure would be the mathematical catenary. An arch built with a thickness just able to contain the catenary would be precariously stable, whereas an arch of double that thickness would easily accommodate a wide range of possible lines of thrust. In practice, a geometrical factor of safety of 2 appears to be appropriate, to allow for building irregularities and for movements imposed by the environment, both of which can distort the original designed geometry.

A slightly more realistic example is shown in fig. 2.24. For small-span (perhaps medieval) bridges, the most critical loading often results from the passage of a single vehicle axle, applied somewhere near quarter-span. The value of P is

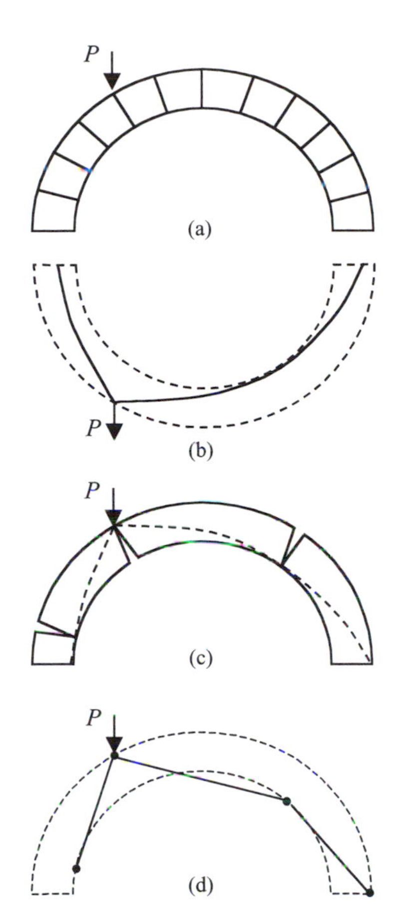

Figure 2.24. (a) Masonry arch under point load, (b) line of thrust, reflected, (c) collapse of arch, (d) corresponding collapse mechanism.

required which just causes the arch to collapse. Figure 2.24(b) shows Hooke's hanging chain, where due allowance has been made for the weight of the arch material and of the fill supporting the roadway; as shown, the hanging chain (i.e. the inverted line of thrust) lies comfortably within the profile of the arch. As the value of P is increased, however, a stage is reached at which the line of thrust can only just be contained [fig. 2.23(c)]; 4 hinges have formed, and the arch is on the point of collapse as a four-bar chain [fig. 2.23(d)].

If an existing arch were being analysed to determine a maximum safe load *P*, the first step would be to shrink the arch ring by an appropriate geometrical factor (e.g. 2), and it is then an easy matter, on the drawing board or by computer, to determine the value of *P* which would just cause collapse of this reduced arch. This would then be the value of the safe load for the real arch. For the purposes of design rather than analysis, the minimum thickness of the arch ring would be determined for the required value of *P*, and the arch would then be built with twice this thickness (to achieve a geometrical factor of 2).

The example illustrated in fig. 2.24 has been simplified, but it is straightforward to carry out similar analyses for trains of loads, and to determine the most critical positions of such trains.

2.6 The structural state

The state of a structure consists of a set of internal stress resultants which are in equilibrium with given external loading. For a hyperstatic structure the equilibrium state is not unique – there are an infinite number of such states. The actual state is determined in the first place by imperfections of manufacture and assembly, which generate internal self-stresses before any external loading is applied. Second, a real structure, already stressed in this way, is subjected to small (in general, unknown) movements of the environment to which it is attached, and very small changes in these boundary conditions can lead to large changes in the internal stress resultants.

Any scheme of calculation which purports to determine the actual condition of a structure (e.g. internal stresses, deformations) gives only one possible equilibrium state – a state, moreover, which is not observable in practice, since the imperfections and boundary conditions are not in fact known. Nevertheless it has been shown in this Chapter that, provided the structural assumptions are obeyed (i.e. ductility, small deformations), the overall strength of the structure may be calculated with confidence.

The simple hyperstatic arch shown in fig. 2.21 becomes statically determinate when subjected to a small spread of the abutments. Two hinges form at or near the springings, and the central hinge generates the three-pin arch. The exact position of the central hinge depends weakly on the applied (gravity) loading; if this loading is symmetrical about the mid-span of the arch, then the hinge forms exactly at mid-span (under certain circumstances, the hinge may split into two), or otherwise at a nearby joint. In this case, then, the movements of the abutments have determined a physically unique state for the structure, and this state is almost independent of any reasonable loading system.

A hyperstatic steel or concrete structure does not in general become determinate under similar circumstances, but the same general conclusions hold. The state of a structure is determined by imperfections of construction and by deformations imposed by the environment, but the resulting internal-stress resultants must always obey the overall equations of equilibrium. The plastic method of design selects an equilibrium state which leads effectively to the lowest possible stresses throughout the structure; that "plastic" state is also not observable, but the design is safe.

3 Stiffness

3.1 The truss

The statically determinate truss discussed in Chapter 2 is redrawn in fig. 3.1. As before, members AB, BC and CD are assumed to be infinitely strong and stiff. The diagonal bar AC, of cross-sectional area A and Young's modulus E, is the only active member, and its elastic extension due to the force it is carrying is

$$e_{\mathrm{AC}} = \frac{(W\sqrt{2})\,(l\sqrt{2})}{AE} = 2\frac{Wl}{AE}. \tag{3.1}$$

As was seen from the displacement diagram of fig. 2.10, the horizontal deflexion of the loading point C is therefore

$$\Delta = 2\sqrt{2}\frac{Wl}{AE}. \tag{3.2}$$

It was previously noted that the construction of displacement diagrams is clumsy and difficult; a more elegant way of deriving deflexions is through the use of virtual work.

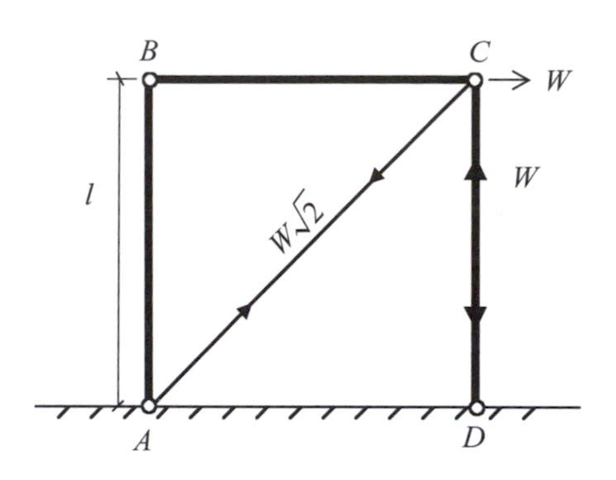

Figure 3.1. Statically determinate truss.

For trusses, it is far easier to write equilibrium equations than to establish relations between bar extensions and corresponding joint displacements. Thus, in using virtual work, the computed bar extensions, such as eq. (3.1), together with the joint displacements Δ (to be determined), are taken as the compatible set. On the other hand, the equilibrium statement entering the virtual-work eq. (2.3) is now a set of bar forces which equilibrate a unit load acting on the truss; the unit load is applied at the point at which it is required to calculate the deflexion. For the simple statically determinate example, if the horizontal deflexion of joint C is to be determined, then a unit horizontal load is applied at that joint (fig. 3.2). Figure 3.2 also shows the bar forces in equilibrium with the unit load, and the information may be tabulated – see Table 3.1.

Thus, writing eq. (2.3) in full,

$$1.\Delta = (0)\,(0) + (0)\,(0) + (-1)\,(0) + (\sqrt{2})\,\left(2\frac{Wl}{AE}\right), \quad (3.3)$$

which is, of course, eq. (3.2). The dummy unit load has identified the displacement Δ whose value was required.

Table 3.1

	AB	BC	CD	AC
Actual forces due to W (fig. 3.1)	0	0	$-W$	$W\sqrt{2}$
Actual elastic bar extensions e	0	0	0	$2Wl/AE$
Bar forces due to unit load (fig. 3.2)	0	0	-1	$\sqrt{2}$

As for the analysis of forces, the calculation is trivial for a statically determinate structure. However, the technique is of great power when applied to the hyperstatic truss shown in fig. 2.4, redrawn as fig. 3.3(a). As before, the two structural members AC and BD are taken as identical, with cross-sectional area A and Young's modulus E, and capable of carrying both tension and compression; the truss is supposed to be initially stress-free. It was shown that only one relationship, equation (2.1), could be established from the use of statics, namely

$$P_1 - P_2 = \sqrt{2}W. \tag{3.4}$$

A second equation for the elastic state of the truss can be found only by using the other two structural equations;

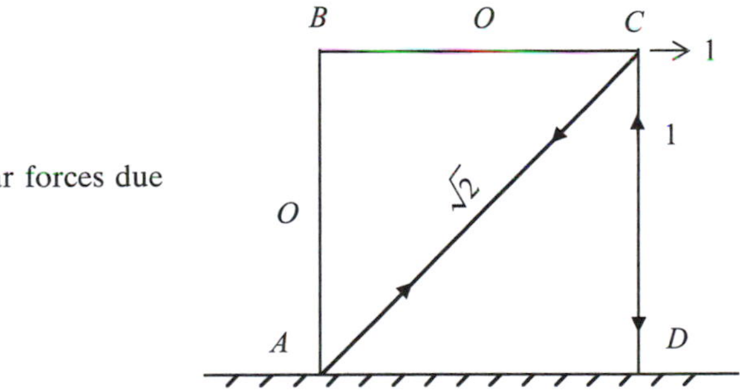

Figure 3.2. Bar forces due to unit load.

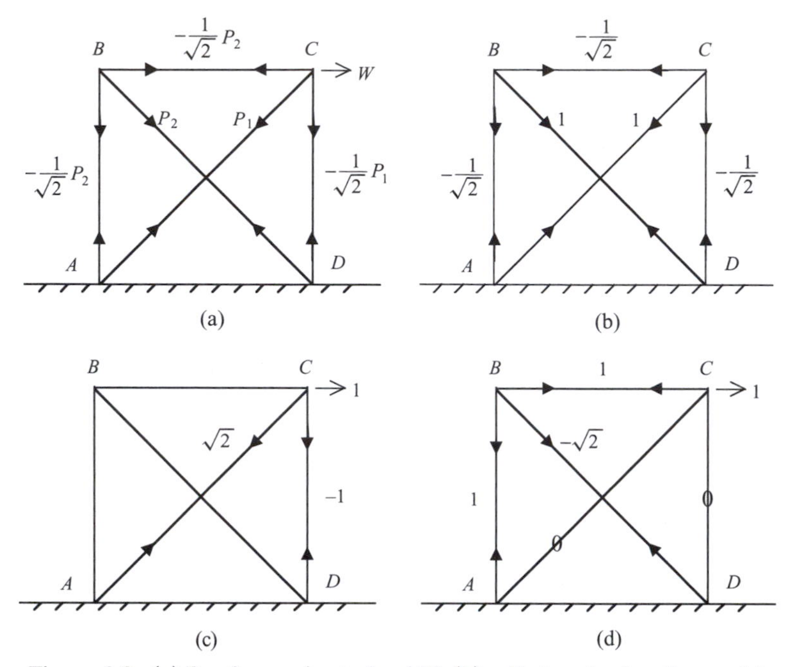

Figure 3.3. (a) Bar forces due to load W, (b) self-stressing bar forces, (c) and (d) two possible sets of bar forces equilibrating unit load.

the elastic properties of the bars must be introduced, and the deformation of the truss must be considered. The actual bar extensions corresponding to the bar forces of fig. 3.3(a) are shown in Table 3.2, together with additional information.

Since the three outer bars are rigid, and suffer no extension, the first 3 columns of the table could have been omitted.

The second line of the table gives the self-stressing bar forces shown in fig. 3.3(b), and these are used, together with the actual bar extensions in the first line of Table 3.2, in the

Table 3.2

	AB	BC	CD	AC	BD
Actual elastic bar extensions	0	0	0	$\frac{\sqrt{2}P_1 l}{AE}$	$\frac{\sqrt{2}P_2 l}{AE}$
Self-stressing forces [fig. 3.3(b)]	$-\frac{1}{\sqrt{2}}$	$-\frac{1}{\sqrt{2}}$	$-\frac{1}{\sqrt{2}}$	1	1
Bar forces due to unit load [fig. 3.3(c)]	0	0	–1	$\sqrt{2}$	0
Bar forces due to unit load [fig. 3.3(d)]	1	1	0	0	$-\sqrt{2}$

virtual-work eq. (2.3). Since there are no external loads, the left-hand side of the equation is zero and, in full,

$$\left(-\frac{1}{\sqrt{2}}\right)(0)+\left(-\frac{1}{\sqrt{2}}\right)(0)+\left(-\frac{1}{\sqrt{2}}\right)(0)+\left(\frac{\sqrt{2}P_1 l}{AE}\right)(1)+\left(\frac{\sqrt{2}P_2 l}{AE}\right)(1)=0; \tag{3.5}$$

that is,

$$P_1+P_2=0. \tag{3.6}$$

Equations (3.4) and (3.6) solve to give

$$P_1=-P_2=\frac{1}{\sqrt{2}}W, \tag{3.7}$$

and the elastic solution has been obtained. [For a truss with a greater degree of redundancy, then further self-stressing systems must be constructed. For example, the truss in fig. 2.8 has three redundancies and, conveniently, the self-stressing forces of fig. 3.3(b) could be applied in turn to each square panel, leading to three equations similar to eq. (3.5)].

It is now possible to proceed to the calculation of deflexions – for example, the horizontal displacement Δ of the loading point C. As before, a unit dummy load is applied, and fig. 3.3(c) shows a set of bar forces in equilibrium with the unit load. These are certainly not the actual forces produced by the load, but they satisfy the equations of equilibrium at each joint, and this is all that is needed; the forces are displayed in the third line in Table 3.2. Using virtual work as before to combine the first and third lines of the table (and omitting contributions from the three outer bars AB, BC and CD since their extensions are zero), thus:

$$1.\Delta = \left(\sqrt{2}\frac{P_1 l}{AE}\right)(\sqrt{2}) + \left(\sqrt{2}\frac{P_2 l}{AE}\right)(0),$$

$$\text{or} \quad \Delta = 2\frac{P_1 l}{AE} = \sqrt{2}\frac{Wl}{AE}. \tag{3.8}$$

Thus the cross-braced truss in fig. 3.3 is stiffer than the simple truss in fig. 3.1; the addition of the extra member has halved the deflexion Δ.

To reinforce the statement that *any* equilibrium set of forces equilibrating the dummy unit load may be used to derive eq. (3.8), an alternative set (of an infinite number) is shown in fig. 3.3(d), and tabulated in the last line in Table 3.2. Using this equilibrium statement with the elastic bar extensions of the first line in the table:

$$1.\Delta = \left(\sqrt{2}\frac{P_1 l}{AE}\right)(0) + \left(\sqrt{2}\frac{P_2 l}{AE}\right)(-\sqrt{2}),$$

$$\text{or} \quad \Delta = -2\frac{P_2 l}{AE} = \sqrt{2}\frac{Wl}{AE}, \quad \text{as before.} \tag{3.9}$$

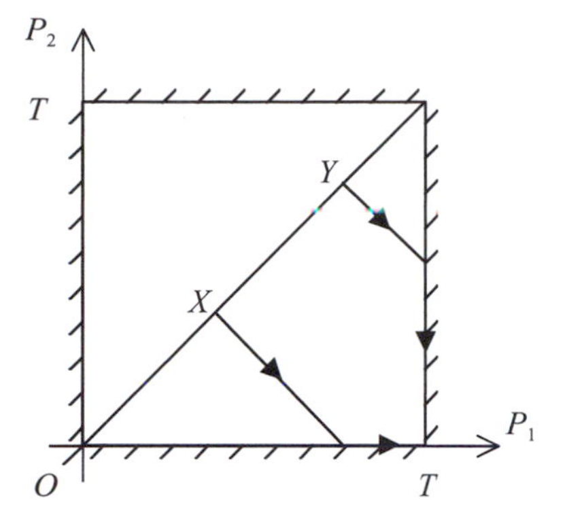

Figure 3.4. Yield surface for truss braced by flexible cables. Initial state of pre-stress is represented by X and Y; in either case, limiting load is reached when $P_1 = T$.

The diagonals AC and BD might be strong flexible cables, introduced for the purpose of stiffening the truss; if they were assembled to be just tight, then application of the load W would immediately cause cable BD to go slack, and the truss would revert to that shown in fig. 3.1. To prevent this, the cables are therefore pre-tensioned (perforce with equal loads, as has been seen). Since all the analytical results have been linear, the solution given [e.g. eq. (3.7)] may be simply superimposed on the pre-existing stress state that is possible for a hyperstatic structure. As the load W is slowly increased, the force in cable BD decreases from its initial tensile value until it reaches zero.

The yield surface shown in fig. 2.7 is redrawn in fig. 3.4, with $C = 0$ since the cables are unable to accept compression. The initial state of pre-stress is represented by some point X on the line $P_1 = P_2$, and loading from X follows a line parallel to $P_1 = -P_2$, eq. (3.7). When the loading point reaches the P_1 axis, cable BD goes slack, and the load may be further increased until the limiting tensile load T is reached in cable AC. In this second phase the truss is half as stiff; the load/

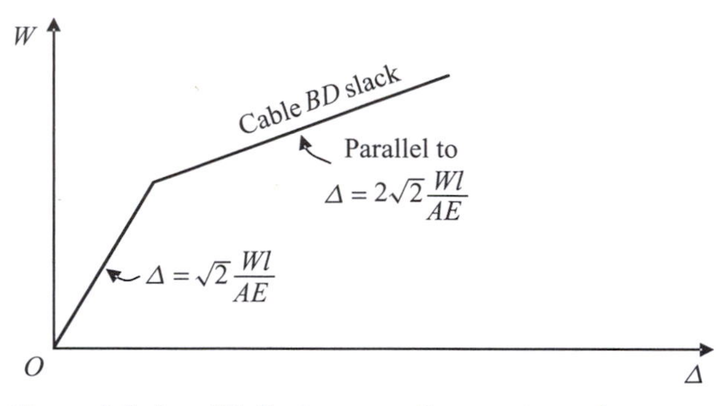

Figure 3.5. Load/deflexion curve for pre-stressed truss.

deflexion curve is shown in fig. 3.5. Alternatively, if the initial tensions were greater than $^1/_2 T$, the starting point would be represented by the point Y in fig. 3.4. Loading causes cable AC to yield while cable BD is still in tension, and the truss can accept further load until this cable goes slack. The load/deflexion curve is identical to that sketched in fig. 3.5, with the legend Cable BD slack replaced by Cable AC yielding.

3.2 Bending stiffness

The equilibrium equations for beams and frames can be represented by bending-moment diagrams (e.g. figs 2.11 and 2.12). As discussed, the bending moments for a hyperstatic beam or frame are not determinable by the equations of statics alone – just as for the truss, material properties and compatibility conditions must be introduced in order to obtain solutions.

For an elastic (prismatic) member in bending, the fundamental equation expressing deformation of that member is

$$M = EI\kappa; \tag{3.10}$$

that is, the curvature κ of the member at any section is proportional to the bending moment M acting at that section. The constant EI can be derived from a (slightly approximate) stress analysis; E is the value of Young's modulus, and I is the second moment of area of the section about the axis of bending. (Values of I for standard beam and column sections are given in the section tables used by the designer; for present purposes, it is the linear relationship between moment and curvature that is important.)

Displacements are assumed to be extremely small, so that the full mathematical expression for curvature,

$$\kappa = \frac{\frac{\mathrm{d}^2 y}{\mathrm{d}x^2}}{\left[1 + \left(\frac{\mathrm{d}y}{\mathrm{d}x}\right)^2\right]^{3/2}}, \tag{3.11}$$

can be replaced by

$$\kappa = \frac{\mathrm{d}^2 y}{\mathrm{d}x^2}, \tag{3.12}$$

since the square of the slope $(\mathrm{d}y/\mathrm{d}x)$ is very small indeed. Thus the basic bending eq. (3.10) becomes

$$M = EI\frac{\mathrm{d}^2 y}{\mathrm{d}x^2}. \tag{3.13}$$

As a simple example, the cantilever beam shown in fig. 3.6 is examined; a point load W is applied at the tip. All the external forces are shown in the figure, including the bending

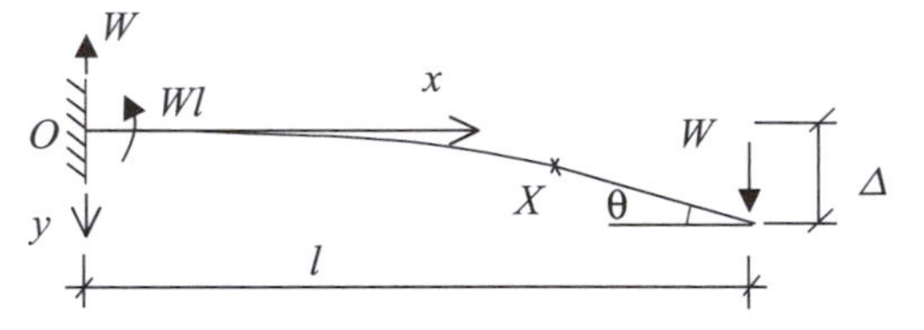

Figure 3.6. Elastic cantilever beam with tip load W.

moment Wl induced at the clamped end, so that the bending moment at any section X can be written

$$M_{\mathrm{X}} = Wl - Wx. \tag{3.14}$$

Equation (3.2) becomes

$$\left.\begin{aligned} EI\frac{\mathrm{d}^2 y}{\mathrm{d}x^2} &= Wl - Wx \\ \text{or}\quad EI\frac{\mathrm{d}y}{\mathrm{d}x} &= Wlx - \frac{1}{2}Wx^2 + A \\ \text{and}\quad EI\,y &= \frac{1}{2}Wlx^2 - \frac{1}{6}Wx^3 + Ax + B, \end{aligned}\right\} \tag{3.15}$$

where A and B are constants of integration.

The third and final statement of structural theory must be used to evaluate these two constants; the compatibility condition emerges as boundary conditions on the differential equation. Figure 3.6 shows that both slope and deflexion are zero at the end $x = 0$, so that, from eqs (3.15), $A = B = 0$. Finally, then, the shape of the deflected elastic cantilever has been found; in particular, the deflexion and slope at the end $x = l$ are given by

$$\left.\begin{aligned} \Delta &= \frac{1}{3}\frac{Wl^3}{EI} \\ \theta &= \frac{1}{2}\frac{Wl^2}{EI}, \end{aligned}\right\} \tag{3.16}$$

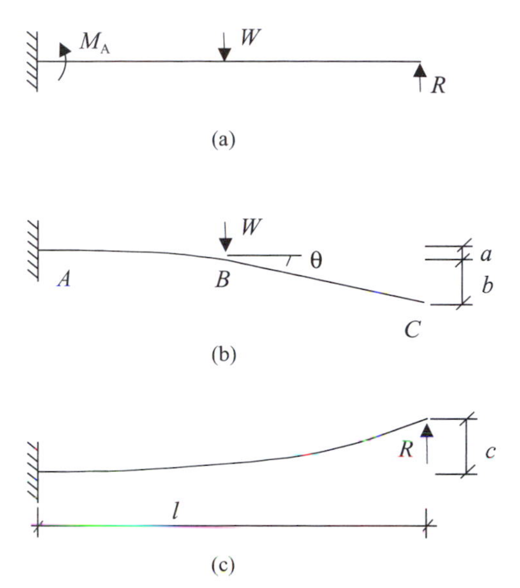

Figure 3.7. (a) Propped cantilever, (b) prop removed, (c) prop force only acting on cantilever.

while the deflexion at mid-span is $5Wl^3/48EI$. These results may be used to find the elastic deflected form of the (hyperstatic) propped cantilever shown in fig. 2.12 and redrawn in fig. 3.7.

The application of the load W at mid-span of the (initially stress-free) beam induces a reaction R at the prop, and the value of R cannot be determined by statics. It may be regarded as the single redundancy in the system, as an alternative to the unknown M_A in fig. 2.12. If the prop were removed, then the (now statically determinate) cantilever would deflect as shown in fig. 3.7(b), with portion AB taking up the bent form indicated by eqs (3.15), and portion BC remaining

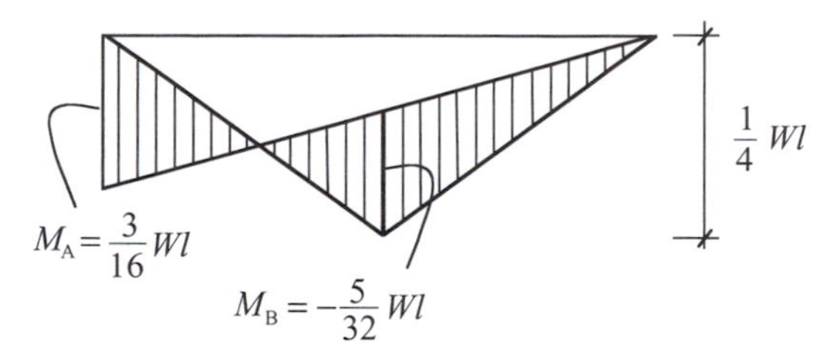

Figure 3.8. Conventional elastic solution for propped cantilever.

straight. The total tip deflexion is found by using the results eqs (3.16):

$$a + b = \frac{1}{3EI}(W)\left(\frac{l}{2}\right)^3 + \left(\frac{l}{2}\right)\left(\frac{1}{2EI}\right)(W)\left(\frac{l}{2}\right)^2 = \frac{5}{48}\frac{Wl^3}{EI}. \tag{3.17}$$

The value of R must be such that the deflexion c in fig. 3.7(c) is exactly equal to the value e.g. (3.17), that is,

$$c = a + b = \frac{Rl^3}{3EI} = \frac{5}{48}\frac{Wl^3}{EI}, \tag{3.18}$$

so that,

$$R = \frac{5}{16}W. \tag{3.19}$$

Thus the value of M_A in fig. 3.7(a) is given by

$$M_A = \frac{Wl}{2} - Rl = \frac{3}{16}Wl, \tag{3.20}$$

and the elastic bending-moment diagram may be constructed as shown in fig. 3.8. The deflexion Δ at the loading point (mid-span) can be evaluated from fig. 3.7 using the value of R that has been determined:

$$\Delta = \frac{Wl^3}{24EI} - \frac{5}{48}\frac{Rl^3}{EI} = \frac{7}{768}\frac{Wl^3}{EI}, \tag{3.21}$$

The corresponding deflexion of a simply supported beam is

$$\Delta = \frac{Wl^3}{48EI}; \tag{3.22}$$

the clamp at the left-hand end has reduced the deflexion in the ratio 7:16.

The deflexion at mid-span of the free cantilever in fig. 3.6, due to a load applied at the tip, was $5Wl^3/48EI$. It is no accident that this is exactly the value of the deflexion of the cantilever at the tip when the load is applied at mid-span [see (fig. 3.7(b)] and is an example of Maxwell's Reciprocal Theorem, which is discussed in Chapter 4.

Figure 3.8 purports to represent the actual elastic state of the propped cantilever beam under the action of the applied load W. However, just as for the truss, any small defects of manufacture, or of assembly, result in an initial state of self-stress, so that the actual values of M_A and M_B differ from those shown in the figure. For example, if the prop at the right-hand end of the beam were set slightly too high, then an initial negative moment would act at end A before any external load was applied, and the total moment acting at A under the application of the load would be less than the value shown in fig. 3.8. Correspondingly, the value of M_B would be greater.

The derived values of M_A and M_B are in fact the *changes* in value of bending moment due to the applied load, and must be superimposed on the (unknown) initial state of self-stressing moments.

However, the computed values of deflexion, such as that in eq. (3.21), give an estimate of the actual elastic response of the structure to its applied loading. That is, a hyperstatic structure as built has defects and its initial stress state is, in general, unknown, but the structure may well respond to loading in a way calculable by classical elastic theory.

Unfortunately, not all defects are of the sort exemplified by a rigid prop being set, initially, at the wrong level. For example, the elastic solution of the propped cantilever required the use of the condition of zero slope of the beam at the clamped end. Just as the rigid prop could in fact give way, perhaps elastically, under the action of load, so a supposedly clamped end could rotate as bending moments are developed in the beam. If such a rotation did take place, then the value of M_A in fig. 3.8 would be reduced, and the numerical value of M_B correspondingly increased; the value of deflexion in eq. (3.21) would also be increased.

In general, the assumptions of perfection built into an elastic analysis lead to calculated values of deflexions which are likely to be exceeded in practice by a real unclad structure. On the other hand, the effects of cladding may well not be taken into account, and cladding could stiffen a bare skeleton.

Putting these considerations aside, and assuming that the bending moments of fig. 3.8 will obtain in practice, then, as the load is increased, a plastic hinge first forms at the clamped

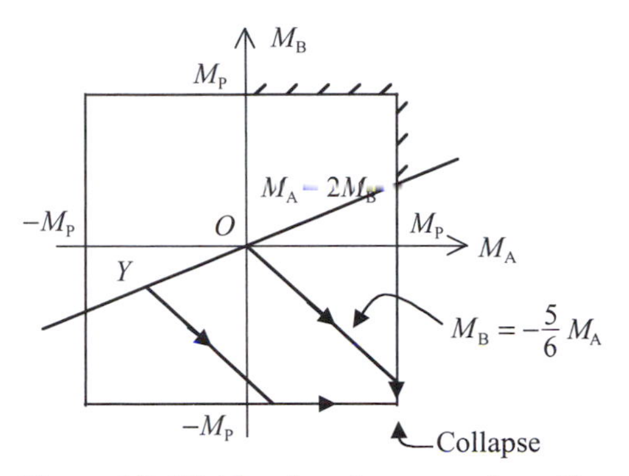

Figure 3.9. Yield surface for propped cantilever.

end A of the beam, when the bending moment there reaches the full plastic value M_P. Thus, starting from the origin O in fig. 3.9 (cf fig. 2.15), the loading path, as the value of W is increased, follows the line $M_B = -\frac{5}{6}M_A$(the values can be verified from fig. 3.8). When this loading path reaches the boundary of the yield surface, a plastic hinge forms at A, but, as usual, the load may be increased until the bending moment at B, under the load, also reaches the full plastic value. During this second stage the stiffness of the propped cantilever is exactly that of the corresponding simply supported beam, since the bending moment at the clamped end A remains constant. The load/deflexion curve is shown in fig. 3.10 (in practice, there will be some rounding at the junctions of the straight lines).

A slightly different situation arises if the beam is initially self-stressed, represented by the point Y in fig. 3.9. The loading path is initially parallel to the line $M_B = -\frac{5}{6}M_A$, but the first plastic hinge now forms at B. In the subsequent stage of

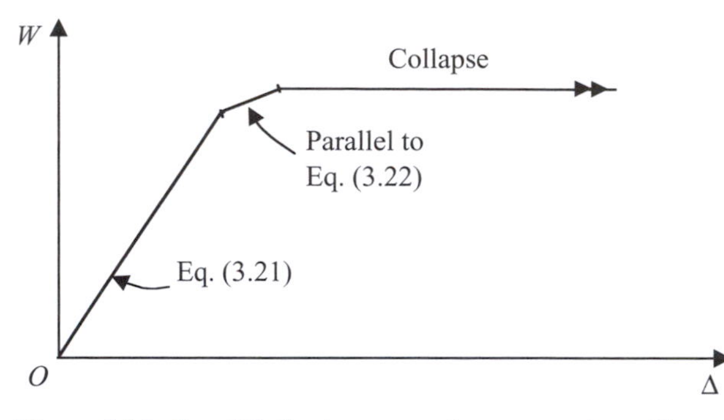

Figure 3.10. Load/deflexion curve for propped cantilever.

loading, the response of the structure is that of a cantilever of length $\frac{1}{2}l$, so that the deflexion, on setting the length to be $\frac{1}{2}l$ in eq. (3.16), is double that of the simply supported beam. The general behaviour is still given schematically in fig. 3.10; however in the second stage the beam is half as stiff.

3.3 Matrix formulation

To estimate the stiffness of a structure – that is, its elastic response to load – somewhat complex calculations are involved. During the course of these calculations values are found for the internal stress resultants (e.g. bar forces, bending moments) and, while knowledge of these may be necessary to check the strength of the members, the final objective is to obtain an equation of the form

$$\Delta = FW. \tag{3.23}$$

An example is given in eq. (3.2), where the (single) deflexion Δ is related to the (single) load W by the factor F, which may be defined as the flexibility of the structure. The internal stress resultants were computed as a necessary step in the analysis, but do not appear specifically in the value of F. Instead, the flexibility is a function of the geometrical properties of the members and the elastic constant(s).

In general, a structure is acted on by a series of loads $W_1, W_2, \ldots$ (i.e. a vector $\mathbf{W}$), and the corresponding elastic displacements of the loading points are $\Delta_1, \Delta_2, \ldots$ (i.e. a vector $\boldsymbol{\Delta}$). (For a three-dimensional truss or frame, each load W is conveniently specified by three components in the direction of the co-ordinate axes, and similarly for each deflexion Δ.) Then, exactly corresponding to eq. (3.23), it is possible to write the general relationship

$$\boldsymbol{\Delta} = \mathbf{FW}, \tag{3.24}$$

where $\mathbf{F}$ is the *flexibility matrix* of the whole structure. If there is a number n of loads (some of which may have zero value) and the same number n of deflexions, then the relationship in eq. (3.24) represents n linear equations. The flexibility matrix $\mathbf{F}$ may be assembled systematically by writing the three basic structural equations; that is:

1. Force/displacement relations are written for individual members (e.g. axial force is related to axial extension, and bending moment to curvature). This is the elastic *stress/strain* relation.
2. Displacements of the joints of the structure are related to deformations of individual members. This is *compatibility*.

3. *Equilibrium* is established throughout the structure; the internal stress resultants are related to the external loads by the equations of statics.

The assembly of the flexibility matrix by computer is straightforward, and very little information is required to calculate a particular truss or frame. The members must be described; that is, their lengths, areas, second moments of area, inclinations, and elastic moduli must be given, and the connexions between members (and the environment) must be specified. The loading system must be defined, although numerical values of the loads are not required at this stage. Nothing else is needed for a properly programmed computer, which eliminates the (for the moment) unwanted internal stress resultants. The resulting flexibility matrix **F** is square (and also symmetric, as a consequence of the reciprocal theorem described in Chapter 4), and it may be inverted to give

$$\mathbf{W} = \mathbf{F}^{-1}\boldsymbol{\Delta} = \mathbf{K}\boldsymbol{\Delta}, \tag{3.25}$$

where **K** is known as the *stiffness matrix*.

The formulation of structural problems in terms of matrices gives a compact and correct way of deriving the elastic state of a given structure under given loading. Moreover, once the equations have been solved for particular values of the loads, back substitution furnishes the values of the internal stress resultants, so that checks on strength and stability may be made. However, the number of equations to be solved for any practical structure of modest size is very large, and exact solutions are really only possible through the use of a high-speed computer. Even then, some economy may be achieved by giving attention to the shape and ordering of the matrices involved.

For nearly two centuries, and before the advent of the computer, ways were sought to reduce the numerical work in the analysis of elastic structures. On the one hand, basic properties of elastic systems were discovered which enabled some shortcuts to be made and, on the other hand, simpler but approximate solutions to the equations were derived. Some of these ideas are discussed in Chapter 4.

4 Elastic analysis

The estimation of the stiffness of a structure, which is essentially an elastic property, involves the calculation of the actual internal stress resultants. Added to the uncertainties engendered by unknown errors of manufacturing and assembly are uncertainties of the elastic properties of the materials used in the structure. Young's modulus for structural steel is known reasonably accurately, but the corresponding modulus for reinforced concrete departs in practice from the value assumed by the designer, may vary from section to section of the structure, and certainly changes with time. The values of internal forces (e.g. bar forces, bending moments) in a computer output must be viewed with some reserve; they are certainly not the values that should be used for strength calculations. [As a simple example, elastic calculations for the propped cantilever in fig. 2.12 show that a displacement of the rigid prop of 1/10 per cent of the span (i.e. 10 mm for a span of 10 m) leads in a typical case to an increase in bending moment of 15 per cent.)]

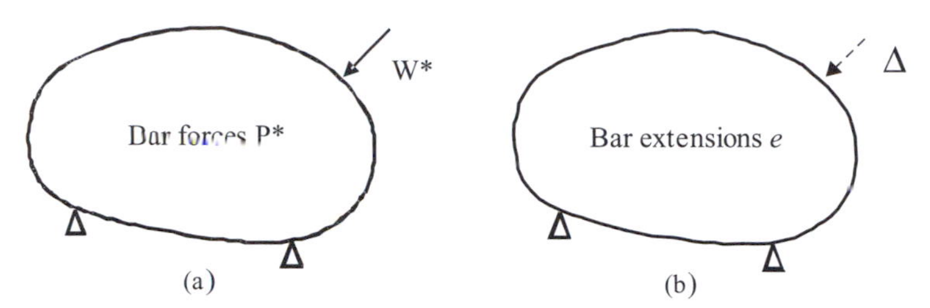

Figure 4.1. General elastic structure: (a) forces in equilibrium, (b) compatible displacements.

As discussed in Chapter 2, elastic parameters need play no part in design for strength, and plastic theory employs ductile yield strength as the only material property. However, plastic methods make no estimate of deflexions – the stiffness of a structure, however it may be affected in practice by unknown considerations, can only be estimated by an elastic analysis. Further, such an analysis furnishes some estimate, however suspect, of the values of the internal forces, which can then be used to check the stability of the members – a topic which again is ignored by a simple plastic approach.

4.1 Elastic properties

Reciprocal theorems

For well over a century, elastic systems have been known to possess some fundamental properties, which can be demonstrated by the application of the principle of virtual work. Figure 4.1 represents an elastic truss (although it could be any elastic structure – for example, a beam or frame resisting loads by bending). Figure 4.1(a) indicates that internal bar forces P^*

are in equilibrium with external loads W^*, while, as a separate matter, displacements Δ are shown in fig. 4.1(b) to be compatible with bar extensions e. Then the equation of virtual work relates these two statements:

$$\sum W^* \cdot \Delta = \sum P^* e. \tag{4.1}$$

Now the bar forces P^* give rise to elastic bar extensions e^*, where

$$e^* = \frac{P^* l}{AE}, \tag{4.2}$$

a typical bar having length l and cross-sectional area A, with E as Young's modulus. Equation (4.1) can therefore be written as

$$\sum W^* \cdot \Delta = \sum \frac{AE}{l} e^* e. \tag{4.3}$$

The actual bar extensions e^* and corresponding deflections Δ^* are associated with the equilibrium system of fig. 4.1(a). Similar loads W and associated bar forces P may be imagined to lead to the deformations of fig. 4.1(b). If, therefore, the roles of figs 4.1(a) and (b) are interchanged, then

$$\sum W \cdot \Delta^* = \sum \frac{AE}{l} e e^*. \tag{4.4}$$

When eqs (4.3) and (4.4) are compared, it is evident that

$$\sum W \cdot \Delta^* = \sum W^* \cdot \Delta. \tag{4.5}$$

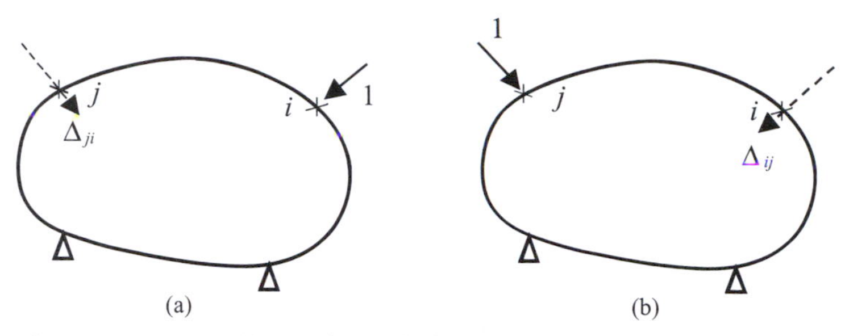

Figure 4.2. Maxwell's Reciprocal Theorem: $\Delta_{ij} = \Delta_{ji}$.

Equation (4.5) was derived for trusses, but it holds for any elastic system, and it is known as Betti's Reciprocal Theorem. In words, if two actual states (i.e. starred and unstarred) of an elastic body are considered, then the work done by the loads W^* of the first state on the displacements Δ of the second state is equal to the corresponding work done by W on Δ^*. It is clear from the derivation that the theorem applies to hyperstatic structures.

Maxwell had stated a few years earlier, in 1864, a more specialized reciprocal theorem. In fig 4.2(a) an elastic structure is subjected to a unit load applied in a specified direction at a location i. As a consequence, the structure deflects through a distance Δ_{ji} in a specified direction at location j of the structure. Similarly, an elastic deflection Δ_{ij} results at location i from the application of a unit load at location j, shown in fig. 3.2(b). Inserting these two statements into eq. (4.5), gives Maxwell's Reciprocal Theorem at once:

$$\Delta_{ij} = \Delta_{ji}, \tag{4.6}$$

and this is effectively a statement of the symmetry of the matrix of elastic flexibility coefficients.

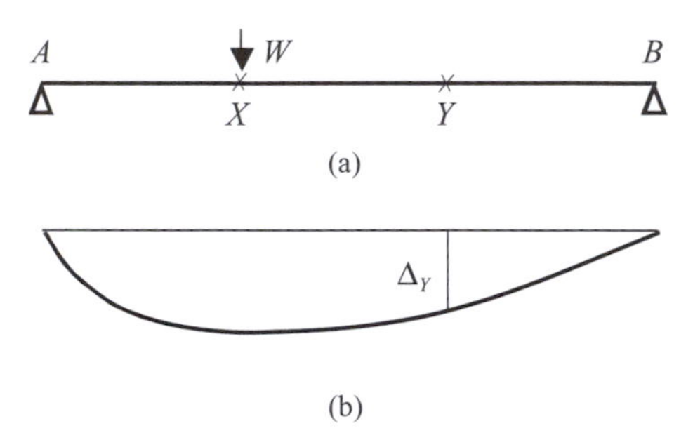

Figure 4.3. Influence line for a simply supported beam.

Influence lines

The simply supported beam shown in fig. 4.3 is subjected to a point load W at some point X within the span; as a consequence an elastic deflexion Δ_Y is observed at some other point Y. Then eq. (4.5) states that if the same load W were applied at Y the same deflexion Δ_Y would be observed at X. This is a special statically-determinate application of the reciprocal theorem. The deflected shape of the beam, fig. 4.3(b), is of interest; it is, in effect, a graph of the deflexion at point X as the load W crosses the beam, and is known as the influence line, usually plotted for a unit value of the load W.

Figure 4.4 shows a simply supported beam with an additional internal support, the system thus being hyperstatic with one redundancy; the supports are supposed to be able to resist both upwards and downwards loads. The reactions due to the applied load W are denoted A, B and C in fig. 4.4(a), and it is required to find one of these, say C. In fig. 4.4(b), a small (virtual) displacement Δ_C^* of the supposedly rigid support at C has been imposed on the otherwise unloaded beam, inducing

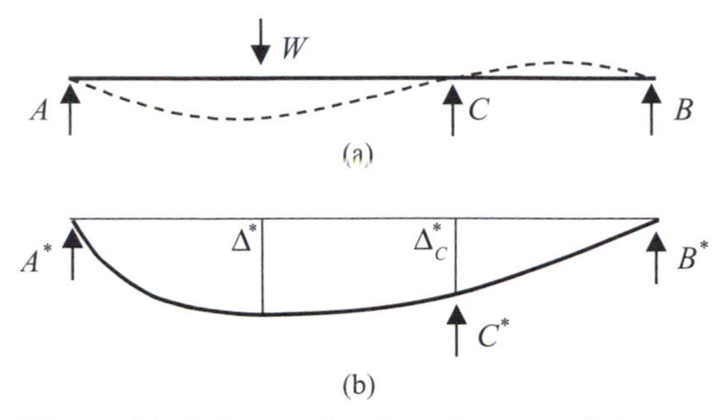

Figure 4.4. Influence line for a hyperstatic beam.

reactions A^*, B^* and C^* at the supports. If eq. (4.5) is applied to the two states sketched in fig. 4.4, then

$$(A)(0) + (W)(\Delta^*) + (C)(-\Delta^*_C) + (B)(0) \\ = (A^*)(0) + (C^*)(0) + (B^*)(0); \quad (4.7)$$

that is,

$$C = \frac{\Delta^*}{\Delta^*_C} W. \quad (4.8)$$

Thus, for an arbitrary unit displacement $\Delta^*_C = 1$ of the internal support, and for a unit load W, the value of the reaction C at the internal support is equal to Δ^*. Once again fig. 4.4(b) gives, to some scale, the influence line for the statically indeterminate reaction C as a unit load W crosses the span.

These ideas can be used to determine internal forces in a frame. In fig. 4.5, for example, an imaginary arbitrary (unit) kink is introduced at the internal support to determine the bending moment M in the beam at that support which results from the application of the load W. The deflected form of the

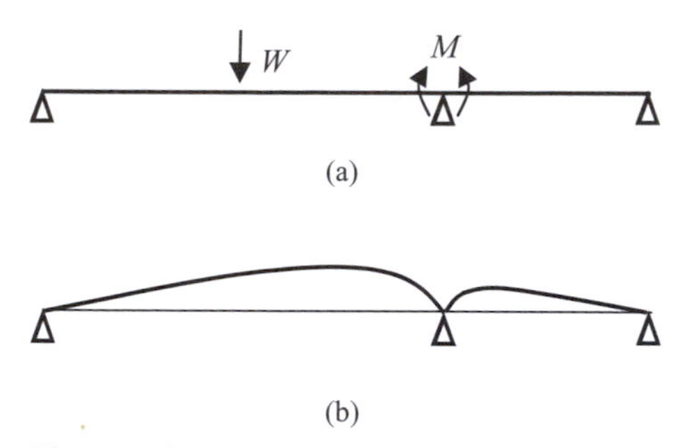

Figure 4.5. Influence line for an internal bending moment.

beam shown in fig. 4.5(b) gives the influence line, and is a plot to some scale of the value of the bending moment M as the (unit) load W crosses the beam.

Indirect model tests

The right-hand side of eq. (4.7) is zero by virtue of the fact that the supports in the original real beam of fig. 4.4 are rigid. Indeed, any deformation similar to that shown in fig. 4.4(b), for which imaginary displacements are introduced at supports which are in fact rigid, or similar to that in fig. 4.5(b), in which an imaginary internal dislocation is imposed, leads to an equation of the form

$$\sum W \cdot \Delta^* = 0, \tag{4.9}$$

[cf eqs (4.5) and (4.7)]. Since eq. (4.9) is homogeneous in the starred deflexion components, it would be possible to make real rather than imaginary displacements on a scale model of the structure. The only requirement is that the scale model of a beam or frame has flexural rigidities that are the same

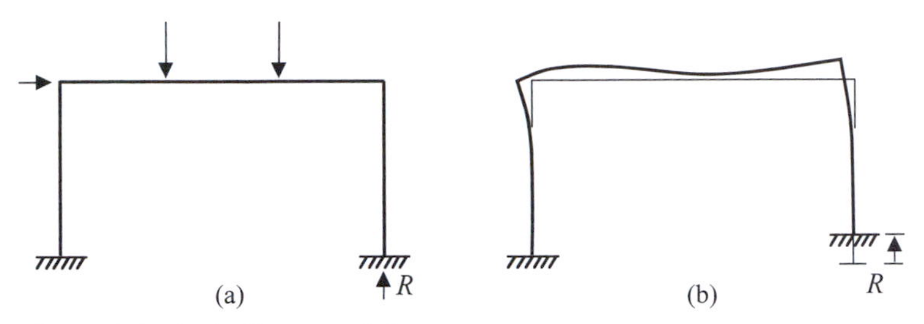

Figure 4.6. An indirect model test.

constant proportion from section to section as those of the original.

Beggs's deformeter makes use of this idea. Real deformations are imposed on a carefully made and properly scaled celluloid model. (Beggs originally used cardboard, which can give acceptable observations.) The model is cut from a sheet of uniform thickness, the depths of the members being varied to ensure correct values of the flexural rigidities. The required deflexion coefficients can then be found experimentally. For example, the portal frame shown in fig. 4.6(a) is subjected to specified loading, and it is required to find the value of R, the vertical component of the force induced at the foot of the right-hand column. The corresponding column foot in the model is given a small vertical displacement, without rotation [fig. 4.6(b)], and the observed distorted shape of the frame provides the deflexion coefficients at the points where the loads are applied to the real structure.

Energy methods

The structural energy theorems are associated with Castigliano, who developed them with reference to trussed frameworks (although he extended them to beams and frames, and indeed to masonry arches). The essential concept is that of internal strain energy, U. For example, if external loads $W_1, W_2 \ldots$, or, in general W_i, induce elastic bar forces P in the bars of a truss, then the internal strain energy U is given by

$$U = \frac{1}{2} \sum Pe = \frac{1}{2} \sum \frac{P^2 l}{AE}, \tag{4.10}$$

where, as usual, l and A are the length and area of a particular bar, and E is Young's modulus. Then Castigliano's first theorem, part 1, is

$$\frac{\partial U}{\partial \Delta_i} = W_i, \tag{4.11}$$

and part 2 of the same theorem is

$$\frac{\partial U}{\partial W_i} = \Delta_i, \tag{4.12}$$

where Δ_i is the displacement of the truss at the section where W_i is applied.

Castigliano's second theorem is concerned with evaluating the redundancies in a hyperstatic structure. For a truss, if there are several redundant bars, and the forces in these bars are, for example, $R, S, T \ldots$, then the values of those forces are such as to make the strain energy U a minimum, that is

$$\frac{\partial U}{\partial R} = \frac{\partial U}{\partial S} = \frac{\partial U}{\partial T} = \quad \cdots \quad = 0. \tag{4.13}$$

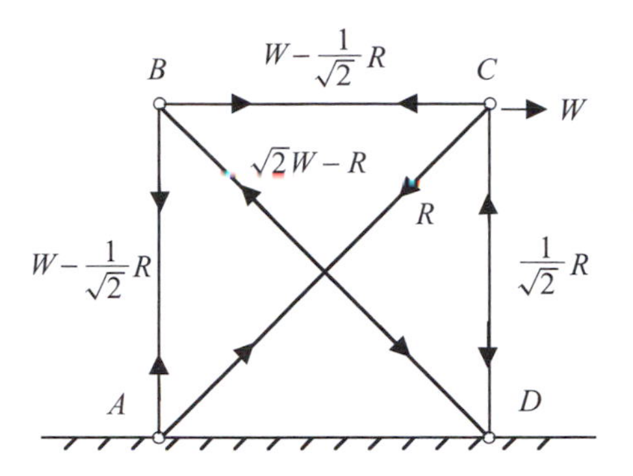

Figure 4.7. The hyperstatic truss.

There are, of course, exactly as many equations (i.e. eq. 4.13) as there are unknown (i.e. redundant) quantities whose values are to be found.

The theorem may be applied to the evaluation of the simple truss of one redundancy, the strength of which was explored in Chapter 2 and its stiffness in Chapter 3. Figure 3.3 is redrawn with a slight change of notation in fig. 4.7; the single redundancy has been labelled R in bar AC. As before, the three outer members are assumed to be rigid, and the active members of the truss are the diagonals, each of length $l\sqrt{2}$. The statics have been completed in fig. 4.7, and diagonal BD carries a compression of $(\sqrt{2}W - R)$. (The strain energy U involves the squares of the bar forces, so that distinguishing by sign between tension and compression is not of immediate importance.) Using eq. (4.10), the strain energy stored internally due to the load W is

$$U = \frac{1}{2}\frac{l\sqrt{2}}{AE}\left[R^2 + \left(\sqrt{2}W - R\right)^2\right]. \tag{4.14}$$

Thus, using Castigliano's second theorem, eq. (4.13),

$$\frac{\partial U}{\partial R} = 0 = R - \left(\sqrt{2}W - R\right), \tag{4.15}$$

or $R = W/\sqrt{2}$ [cf eq. (3.7)].

Castigliano's first theorem, part 2, eq. (4.12), may be used to determine the horizontal deflection Δ of the loading point C:

$$\Delta = \frac{\partial U}{\partial W} = \frac{1}{2}\frac{l\sqrt{2}}{AE}\left[2\sqrt{2}(\sqrt{2}W - R)\right],$$

or

$$\Delta = \sqrt{2}\frac{Wl}{AE}, \tag{4.16}$$

on substituting the value of R already found [cf eq. (3.8)]. (In performing the partial differentiation of the strain energy U it is not necessary to enter the value of R; it may be treated as if it were a constant.)

The proofs of Castigliano's theorems are straightforward by the use of virtual work. For example, if the second theorem were used to determine the values $R, S, T \dots$ of the forces in the redundant bars of a truss, then the first step is to use the equations of statics to compute the forces in all the bars. A typical bar carries a force P, where

$$P = \sum a_i W_i + \alpha R + \beta S + \gamma T + \cdots. \tag{4.17}$$

Here, there are a number of loads W_1, W_2, W_3 . . . acting on the truss, and the coefficients a_i, α, β, γ . . . are simple numbers, such as the $\sqrt{2}$ which occurs in the truss shown in fig. 4.7. The corresponding extension of the typical bar is

$$e = \frac{l}{AE}\left[\sum a_i W_i + \alpha R + \beta S + \gamma T + \cdots\right], \quad (4.18)$$

where the length and cross-sectional area must be inserted appropriately for each bar of the truss. These actual bar extensions e are used as the compatible set in the equation of virtual work.

The equilibrium statement involves self-stressing sets of bar forces, in equilibrium therefore with zero external load. A first such set may be found by setting all but one of the force quantities equal to zero (e.g. say all $W_i = S = T = \cdots 0$), while R is given the value unity. Thus, from eq. (4.17), the self-stressing force in a typical bar is just α, and these forces may be evaluated throughout the truss. Thus, using this force set with the compatible set in eq. (4.18),

$$0 = \sum \alpha e = \sum \frac{l}{AE}\alpha\left[\sum a_i W_i + \alpha R + \beta S + \gamma T + \cdots\right]. \quad (4.19)$$

The full expression for the strain energy, eq. (4.10), is

$$U = \frac{1}{2}\sum \frac{l}{AE}\left[\sum a_i W_i + \alpha R + \beta S + \gamma T + \cdots\right]^2, \quad (4.20)$$

and it is evident that $\partial U/\partial R = 0$ [eq. (4.13)] gives precisely eq. (4.19). The calculation is of course repeated for self-stressing forces induced by $S = 1$, $T = 1$, and so on.

Proofs may be constructed on similar lines for Castigliano's first theorems, eqs (4.11) and (4.12), concerned with deflexions. Should there be no load acting at a particular section at which the deflexion is required, then a dummy load (denoted X) may be applied at that section, and carried through the analysis. Equation (4.12) gives $\partial U/\partial X = \Delta_X$ and, at this stage, the value of X may be set equal to zero.

4.2 Methods of calculation

Castigliano's theorems represent basic properties of elastic systems but, before the advent of the computer, they also provided powerful ways of dealing with the large numbers of equations involved in the analysis of elastic structures. The theorems apply to any elastic system, but are particularly well adapted to the solution of trusses; they become cumbersome when applied to structures in bending. Strain energy in bending (i.e. $M^2/2EI$ per unit length) must be integrated throughout the length of a beam or frame, and the algebraic work can become very heavy (although the partial differentials involved in a strain-energy solution may be performed under the integral sign).

Similarly, the direct solution of the bending equations to determine deflexions involves – as previously discussed – the writing of second-order differential equations. At least some of the work involved can be reduced by making use of standard results. Table 4.1, for example, shows coefficients involved in the behaviour of the simple cantilever; these coefficients were used in Chapter 3 to investigate the elastic response of the hyperstatic propped cantilever (see fig. 3.7).

Slope-deflexion equations

Similar deflexion coefficients can be evaluated for the simply supported beam as recorded in Table 4.2. In this table,

Table 4.1

	End slope	End deflexion
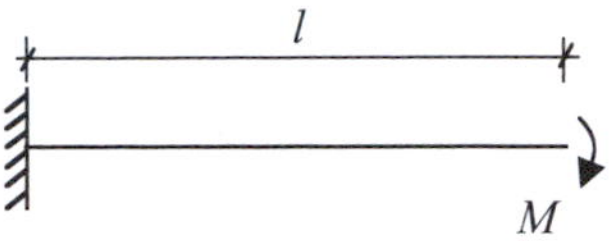 Cantilever, length l, end moment M	$\frac{Ml}{EI}$	$\frac{Ml^2}{2EI}$
Cantilever, end load W	$\frac{Wl^2}{2EI}$	$\frac{Wl^3}{3EI}$
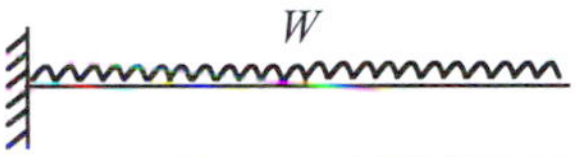 Cantilever, uniformly distributed load W	$\frac{Wl^2}{6EI}$	$\frac{Wl^3}{8EI}$

Table 4.2

	End slope, θ_A	End slope, θ_B
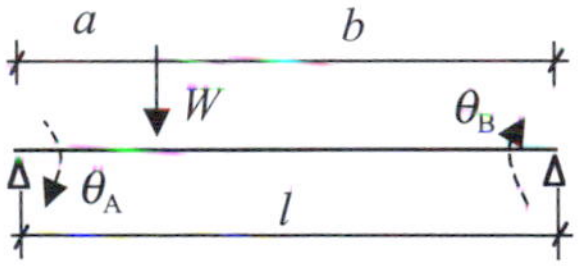 Simply supported, span l, point load W at distances a, b	$\frac{Wab}{6EIl}(l+b)$	$-\frac{Wab}{6EIl}(l+a)$
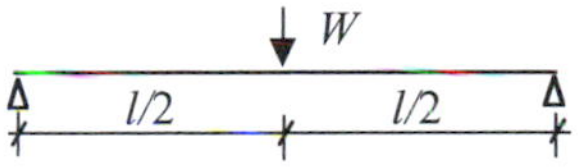 Simply supported, central load W ($l/2$, $l/2$)	$\frac{Wl^2}{16EI}$	$-\frac{Wl^2}{16EI}$
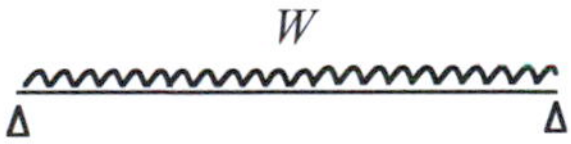 Simply supported, uniformly distributed load W	$\frac{Wl^2}{24EI}$	$-\frac{Wl^2}{24EI}$
 Simply supported, end moment M at B	$-\frac{Ml}{6EI}$	$\frac{Ml}{3EI}$

clockwise end slopes of the beam are denoted as positive. For example, to analyse a continuous beam (i.e. a hyperstatic beam resting on a number of supports), the deflexion coefficients in Table 4.2 may be written individually for each span, and then pieced together to give a number of (linear) simultaneous equations. Further economies in the computational process may be made by replacing specified loading patterns by equivalent fixed-end moments.

Figure 4.8(a) shows an initially straight undeflected member *AB,* which could be part of a continuous beam or of a frame. Under the action of a given system of loads, the member moves to $A'B'$, and experiences clockwise end bending moments M_A and M_B and clockwise end rotations ϕ_A and ϕ_B. The deflexions of the two ends are Δ_A and Δ_B. The general slope-deflexion equations relate these bending moments, rotations and deflexions to the external loading.

If the beam were fixed-ended, the external loading would produce certain moments M_A^F and M_B^F at those ends, which are the fixed-end moments shown in fig. 4.8(b). If now the system shown in fig. 4.8(c) is superimposed on that in fig. 4.8(b), then the original system is recovered; such superposition is possible as a consequence of the linear nature of the elastic equations of bending. The immediate problem is to derive relationships connecting the quantities shown in fig. 4.8(c), which relate to a beam without external loading. First, it is clear that if the beam were moved laterally without bending so that the ends *A* and *B* occupied the positions shown in fig. 4.8(c), then the slope of the beam would be uniform and equal to $(\Delta_B - \Delta_A)/l$. The effect of the end bending moments is to increase the rotation

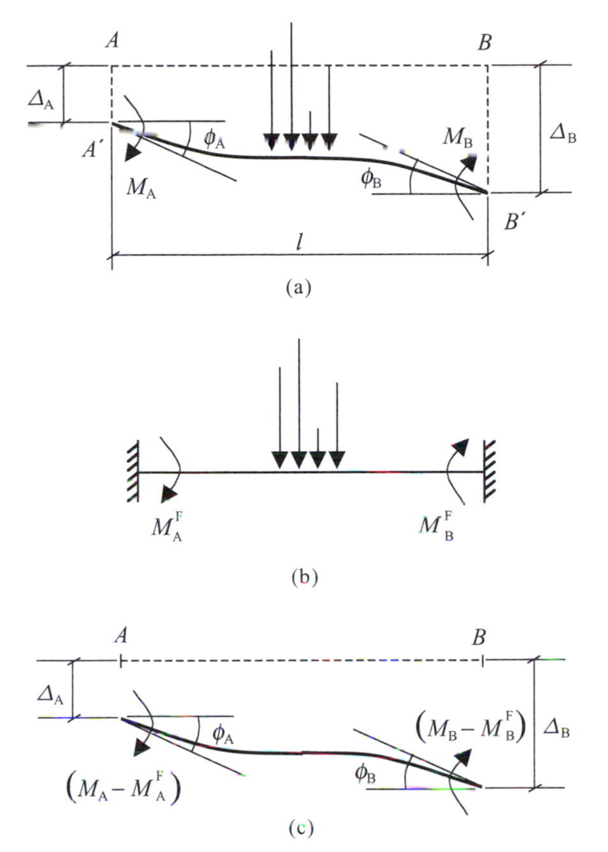

Figure 4.8. Derivation of slope-deflexion equations.

at end A from $(\Delta_B - \Delta_A)/l$ to the final value ϕ_A and, similarly, for end B. Thus, using the final line in Table 4.2,

$$\phi_A - \left(\frac{\Delta_B - \Delta_A}{l}\right) = \frac{(M_A - M_A^F)\,l}{3EI} - \frac{(M_B - M_B^F)\,l}{6EI}, \quad (4.21)$$

$$\phi_B - \left(\frac{\Delta_B - \Delta_A}{l}\right) = \frac{(M_B - M_B^F)\,l}{3EI} - \frac{(M_A - M_A^F)\,l}{6EI}. \quad (4.22)$$

These two equations can be written in the form in which they are normally used:

$$\left.\begin{aligned} \phi_{\mathrm{A}} &= \left(\frac{\Delta_{\mathrm{B}} - \Delta_{\mathrm{A}}}{l}\right) + \frac{l}{6EI}\left\{2\left(M_{\mathrm{A}} - M_{\mathrm{A}}^{\mathrm{F}}\right) - \left(M_{\mathrm{B}} - M_{\mathrm{B}}^{\mathrm{F}}\right)\right\} \\ \phi_{\mathrm{B}} &= \left(\frac{\Delta_{\mathrm{B}} - \Delta_{\mathrm{A}}}{l}\right) + \frac{l}{6EI}\left\{2\left(M_{\mathrm{B}} - M_{\mathrm{B}}^{\mathrm{F}}\right) - \left(M_{\mathrm{A}} - M_{\mathrm{A}}^{\mathrm{F}}\right)\right\}, \end{aligned}\right\} \tag{4.23}$$

and this is the form most convenient for solution of beam and frame problems in terms of unknown (i.e. redundant) forces. If deformation variables (i.e. unknown deflexions and rotations) are taken, however, then the slope-deflexion equations may be written:

$$\left.\begin{aligned} M_{\mathrm{A}} - M_{\mathrm{A}}^{\mathrm{F}} &= \frac{6EI}{l}\left\{\frac{1}{3}\left(2\phi_{\mathrm{A}} + \phi_{\mathrm{B}}\right) - \left(\frac{\Delta_{\mathrm{B}} - \Delta_{\mathrm{A}}}{l}\right)\right\} \\ M_{\mathrm{B}} - M_{\mathrm{B}}^{\mathrm{F}} &= \frac{6EI}{l}\left\{\frac{1}{3}\left(\phi_{\mathrm{A}} + 2\phi_{\mathrm{B}}\right) - \left(\frac{\Delta_{\mathrm{B}} - \Delta_{\mathrm{A}}}{l}\right)\right\}. \end{aligned}\right\} \tag{4.24}$$

Equations (4.23) and (4.24) are simplified examples of building blocks which may be assembled into flexibility and stiffness matrices for any elastic structure. The full equations would be three-dimensional, and also would involve axial loads and consequential axial deformation of the members.

Table 4.3 gives three useful sets of fixed-end moments; by simple superposition or by integration, the results given in the first line of the table can be used to evaluate the results of more complex patterns of loading.

Many other methods have been developed for the solution of structural problems, some of them of great ingenuity. Their use results in substantial savings in the labour of

Table 4.3

	M_A^F	M_B^F
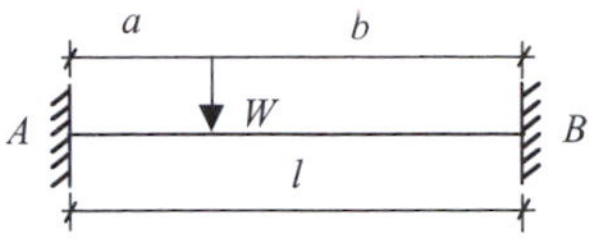 Fixed beam AB, length l, point load W at distance a from A and b from B	$-\dfrac{Wab^2}{l^2}$	$\dfrac{Wa^2b}{l^2}$
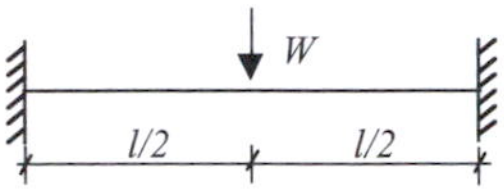 Fixed beam, central point load W ($l/2$, $l/2$)	$-\dfrac{Wl}{8}$	$\dfrac{Wl}{8}$
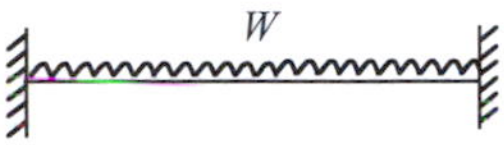 Fixed beam, uniformly distributed load W	$-\dfrac{Wl}{12}$	$\dfrac{Wl}{12}$

computation, but they are in fact merely efficient ways of solving the numerous equations generated by the analysis of even a simple structure, and contribute nothing to basic elastic theory (as do, for example, the theorems of Maxwell and Castigliano). The computational techniques were devised for manual use and are, effectively, unnecessary in the age of the computer.

Almost the last in the long line of techniques, however, which could today be classified as a relaxation method for the solution of linear simultaneous equations, does give some insight into structural behaviour. This is the method of moment distribution, invented by Hardy Cross in 1930.

Moment distribution

Hardy Cross's method starts with an artificial state of compatible deformation for a continuous beam or frame, and then successively adjusts the deformations so that the equations of

Figure 4.9. Moment distribution: the carry-over factor.

equilibrium are satisfied. The artificial state is very simple – all the joints of the structure are assumed to be fixed in position and direction, and the application of the external loads then leads to calculable values of fixed-end moments. These end moments, in general, do not sum to zero at any particular joint; that joint is then unclamped – that is, allowed to rotate so that equilibrium is achieved. Each joint is examined in turn, and in fact re-examined, since rotation at one joint causes extra moments at neighbouring joints. The process, however, is rapidly convergent, and may be carried through to the stage when the required degree of accuracy is attained.

Two basic quantities are needed for the moment-distribution process: *carry-over factors* and *distribution factors*. The propped cantilever shown in fig. 4.9 is subjected to a bending moment M at the pinned end. As a consequence, a bending moment of $\frac{1}{2}M$ is induced at the clamped end, as may be verified immediately by the slope-deflexion eqs (4.23). The carry-over factor for this uniform section member is ½. If the member had a non-uniform section, the carry-over factor would have some other value, depending on the way the cross-section varies.

The same slope-deflexion equations show that the rotation ϕ in fig. 4.9 has the value $Ml/4EI$. Figure 4.10 shows

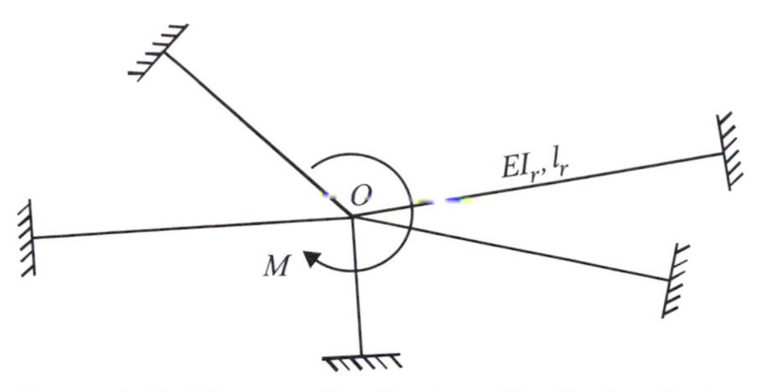

Figure 4.10. Moment distribution: distribution factors.

an assemblage of members of uniform section meeting at a common joint O; the length and flexural rigidity of a typical member are denoted l_r and EI_r. The ends of the members remote from O are all fixed in position and direction, and the ends meeting at O are rigidly connected together. If now a moment M is applied at the joint O, it is *distributed* in some way between the members, with the typical member carrying a moment of value M_r. The joint O may be supposed to rotate through an angle ϕ, imposed equally on all the members; typically,

$$M_r = \frac{4EI_r}{l_r}\phi. \tag{4.25}$$

The sum of all moments induced in the members must be equal to the applied moment M; that is,

$$M = \sum M_r = 4\phi \sum \frac{EI_r}{l_r}. \tag{4.26}$$

Equations (4.25) and (4.26) combine to give

$$M_r = \frac{EI_r/l_r}{\sum EI_r/l_r} M = \frac{k_r}{\sum k_r} M, \tag{4.27}$$

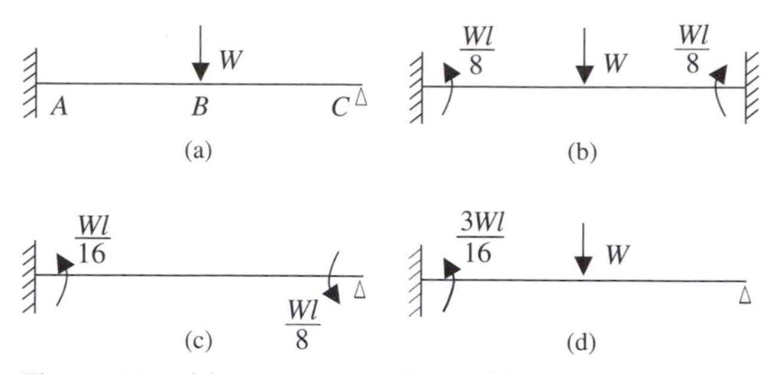

Figure 4.11. (a) Propped cantilever, (b) bending moments with end C clamped, (c) bending moment $Wl/8$ applied at C, (d) superposition of (b) and (c).

where $k_r = EI_r/l_r$ is the *stiffness* of member r. The *distribution factor* is $k_r/\sum k_r$.

This distribution factor was calculated on the assumption that the far ends of all members were fixed against rotation, and it is convenient occasionally to use distribution factors for members whose remote ends are pinned. As shown in Table 4.2, the rotation ϕ corresponding to an applied moment M_r is equal to $M_r l_r/3EI_r$ if the far end of member r is pinned. The effective stiffness of member r is reduced to $\frac{3}{4}EI_r/l_r$.

Moment distribution is a numerical process, but the propped cantilever shown in fig. 2.12 may be used to illustrate the idea of clamping. The beam is redrawn in fig. 4.11(a), and the first step is to clamp the end C and to apply the load W, inducing the fixed-end moments shown in fig. 4.11(b). The clamp at C is now released, since the end is supposed to be pinned – this is equivalent to applying the bending moment $Wl/8$ at end C, as shown in fig. 4.11(c). As a consequence, a moment of half this value is induced at the fixed end A.

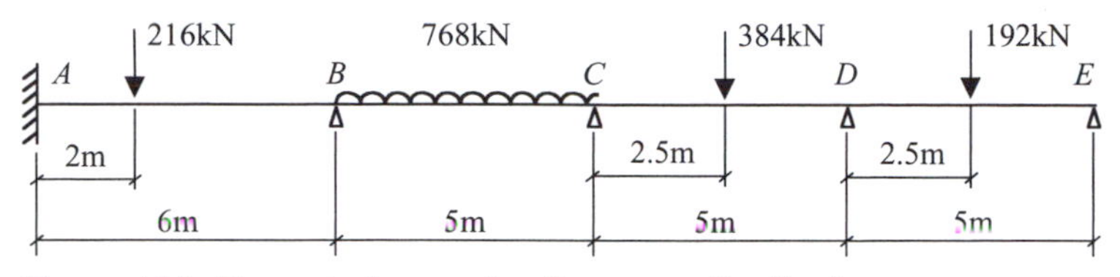

Figure 4.12. Numerical example of moment distribution.

The superposition of figs 4.11(b) and (c) leads to the final state, fig. 4.11(d) (cf fig. 3.8).

The continuous beam in fig. 4.12 has dimensions and carries loads as shown. The section of the beam changes at the supports, and the information required for the solution of the problem is given in Table 4.4. The last line of the table gives the fixed-end moments for each span; they are denoted as positive when they act clockwise on the ends of the beams. These bending moments are *out of balance* at each support; at B, for example, the clockwise moment of 96 kNm in span AB does not balance the anticlockwise moment of

Table 4.4

Span	AB		BC		CD		DE	
Length, m	6		5		5		5	
Flexural rigidity, $B = EI$	$2B_0$		$1.5B_0$		B_0		B_0	
$\frac{B}{l}\left(\times\frac{60}{B_0}\right)$	20		18		12		$\frac{3}{4}(12)$	
Distribution Factors		0.526	0.474	0.6	0.4	0.571	0.429	
Fixed-end moments, kNm	−192	96	−320	320	−240	240	−120	120

320 kNm in span *BC*. The difference of 224 kNm must be provided by some external agency acting at *B*; this external agency does not exist for the actual beam. Thus, to satisfy the condition of equilibrium at the joint, a clockwise moment of 224 kNm must be superimposed there. The calculations can be laid out so that the balancing process can proceed easily and quickly.

Table 4.5 shows the distribution factors and fixed-end moments in lines 1 and 2. Joint *B* is balanced in line 3. The balancing moment of 224 kNm is thought of as being applied to an unloaded structure, with all joints other than *B* (i.e. *A*, *C*, *D* and *E*) remaining clamped. Thus the balancing moment divides between spans *BA* and *BC* according to the distribution factors – that is, 117 kNm to span *BA* and 107 kNm to span *BC*. A line is drawn under the balancing values (line 3 in Table 4.5) to show that joint *B* is now in equilibrium.

The carry-over factor for a uniform beam is $^1/_2$, so that the application of a moment of 117 kNm at end *B* in the span *AB* induces a moment of 59 kNm at end *A*. The two carry-over moments are shown in line 4 of Table 4.5. The bending moments in line 1 of the table correspond to those of a continuous beam with imaginary external moments applied at the supports; the external moments have magnitudes such that the slope of the beam at each support is horizontal. In line 4, the beam remains horizontal at *A*, *C*, *D* and *E*; however, the external moment has been removed from joint *B*, and the beam has been allowed to rotate at this support.

End *A* of the beam is required to remain horizontal, and an external moment can act there. End *E*, however, is pinned,

Table 4.5

Line		A	B		C		D		E
1	Distribution factors		0.526	0.474	0.6	0.4	0.571	0.429	
2	Fixed-end moments	-192	96	-320	320	-240	240	-120	120
3	Balance *B*		117	107					
4	Carry-over, line 3	59			53				
5	Balance *C* and *E*				-80	-53			-120
6	Carry-over, line 5			-40			-27	-60	
7	Balance *B* and *D*		21	19			-19	-14	
8	Carry-over, line 7	11			9	-9			
9	Total bending moments	-122	234	-234	302	-302	194	-194	0

with no external moment. In line 5 of Table 5.4, a bending moment of −120 kNm is superimposed at *E*, which induces a carry-over moment of −60 kNm at *D*, line 6. The stiffness of span *DE* is entered in Table 4.4 as three-quarters of its actual value; the balancing process at a pinned end needs to be carried out only once, and no further moments require balancing. (There is also no carry-over moment to end *E* in line 8 of Table 4.5.)

Several joints can be balanced simultaneously: in line 5, both joints *C* and *E* are balanced together. With the carry-over moments of line 6, joint *B* is now again out of balance (by 40 kNm); this joint and joint *D* are balanced together in line 7. The carry-over moments at joint *C* in line 8 happen to balance exactly, and no further adjustment of the solution is necessary. Had they not balanced, the process could have been continued, the out-of-balance moments being reduced until their values were insignificant.

The final line in Table 4.5 gives the sums of all the bending moments in each column, and these are the values of the bending moments at the joints. The clockwise positive convention implies that all these bending moments are hogging – That is, the nett value at each joint is zero, showing that exact balance has been achieved. The complete bending-moment diagram for the beam can now be constructed. The free bending moments (i.e. each span treated as simply supported) are shown in fig. 4.13, and the reactant line is positioned by the results of the moment distribution; the nett bending moments are given by the difference between the two.

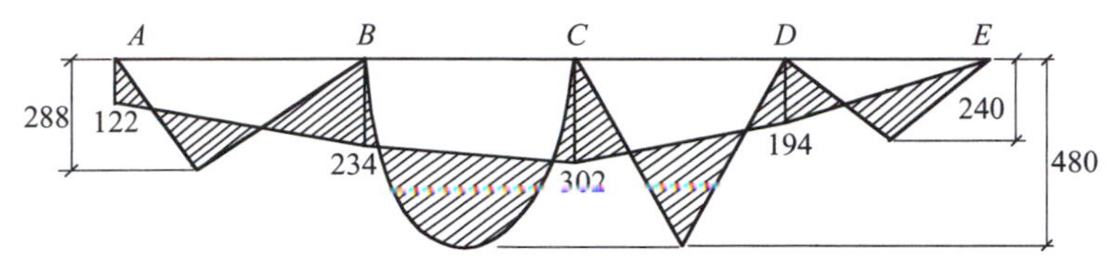

Figure 4.13. Solution of example, fig. 4.12.

Moment distribution may be carried out for frames. For the two-bay, three-storey frame in fig. 2.19(a) there are now 2, 3 or 4 members meeting at each joint, and distribution factors must be calculated accordingly. As before, the joints are first clamped and the external forces are applied, and the resulting bending moments are distributed until balance is attained. However, this process, implicitly, does not allow sway of the frame, and the horizontal loads in fig. 2.19(a) play no part in the analysis, applied as they are to joints assumed to be fixed in position. In reality, the sway of each of the three storeys must be considered. An *elastic* displacement corresponding to the movement in fig. 2.19(c) induces bending moments at the ends of each of the central lengths of the columns, which may be distributed throughout the frame – in all, three such distributions must be made, one for the displacement of each of the three storeys. All these numerical results may then be superimposed so that the governing equilibrium equations are satisfied – in fig. 2.19, the total horizontal force transmitted at each level of the frame is the sum of the horizontal forces applied above that level.

5 Stability

5.1 Elastic buckling

A compressive axial load is applied to an initially straight elastic member (i.e. a column). If the load is truly axial and the member truly straight, then it may be expected that the only observable response is a small shortening (a strain of less than about 1/1000 if the column is of mild steel and is to remain below yield). However, when the load reaches a certain *critical value* the member no longer remains straight, but deflects sideways at a more or less constant value of that load. Analysis of this idealised behaviour illuminates the real structural problem of the buckling of columns.

Figure 5.1(a) illustrates the initially straight uniform member in its supposed deflected state. The column (turned sideways in the figure) has pinned ends, which may approach each other freely to accommodate the axial shortening resulting from the elastic compression and from the development of lateral displacements. Figure 5.1(b) is a free-body diagram of a portion of the column of length x; from overall equilibrium

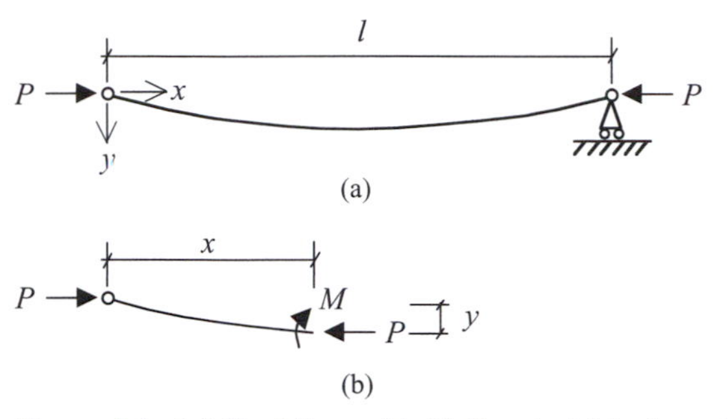

Figure 5.1. (a) Buckling of initially straight pin-ended column under axial load, (b) free-body diagram of part of column.

[fig. 5.1(a)] there can be no shear force in the column, and the only stress resultants acting at the cut are the axial force P and a bending moment M. For equilibrium of the forces in fig. 5.1(b),

$$M + Py = 0, \tag{5.1}$$

and therefore, for the uniform elastic member,

$$EI\frac{\mathrm{d}^2 y}{\mathrm{d}x^2} + Py = 0. \tag{5.2}$$

This is an extremely simple form of the basic differential equation which arises for all similar buckling problems. The solution of eq. (5.2) is

$$y = A\cos\alpha x + B\sin\alpha x, \tag{5.3}$$

where

$$\alpha^2 = \frac{P}{EI}, \tag{5.4}$$

and A and B are constants of integration. At the origin $x = 0$ and $y = 0$, so that $A = 0$. At the other end $y = 0$ at $x = l$, and therefore

$$B \sin \alpha l = 0. \tag{5.5}$$

The solution $B = 0$ is inadmissible, since it contradicts the assumption that the column is in a deflected state, and eq. (5.5) can only be satisfied if

$$\alpha l = 0,\ \pi,\ 2\pi, \ldots \tag{5.6}$$

Again, $\alpha l = 0$ is inadmissible since, from eq. (5.4), this would imply no load on the column; the first meaningful solution to the problem is given by $\alpha l = \pi$, so that

$$P_{\mathrm{e}} = \pi^2 \frac{EI}{l^2}; \tag{5.7}$$

the deflected form is

$$y = B \sin \frac{\pi x}{l}. \tag{5.8}$$

The Euler buckling load is denoted P_{e}; for this particular problem it gives the critical value of axial load at which a pin-ended column buckles. It has not been possible to determine the magnitude of the deflexions – the value of B has not been found and, according to this simple theory, the sinusoidal deflexions of eq. (5.8) can have any (small) magnitude at the same value of P_{e}. Those deflexions are those of a half sine wave; the next solution of the equation, $\alpha l = 2\pi$ from eq. (5.6), represents a full sine wave, and can, theoretically,

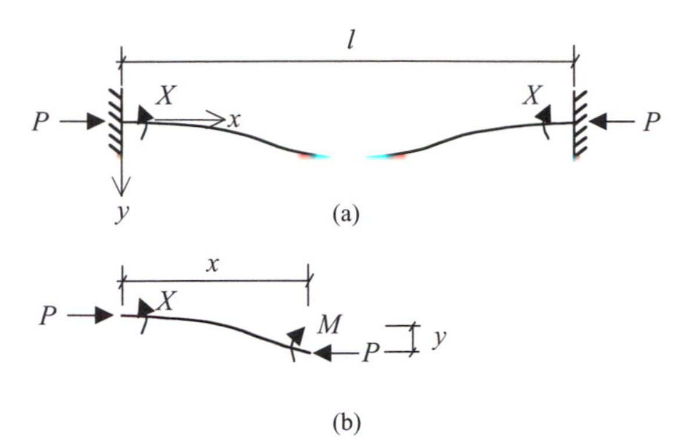

Figure 5.2. (a) Fixed-ended column, (b) free-body diagram.

be maintained in the presence of an axial force of magnitude $4\pi^2EI/l^2$. However, whereas buckling under the Euler load [eq. (5.7)] represents a state of neutral equilibrium, buckling in the higher modes ($\alpha l = 2\pi, 3\pi \ldots$) is highly unstable, and the corresponding loads cannot in practice be attained.

The pin-ended column is statically determinate, and the differential eq. (5.2) can be written immediately. A fixed-ended column is shown, in its supposed buckled state, in fig. 5.2. The two end fixings, while compelling the end to have zero slope, are again free to approach each other axially. It is to be expected that unknown bending moments X are induced at the two ends by the deflexions which are developed, so that the bending moment at a general section, shown in the free-body diagram in fig. 5.2(b), can be written only in terms of X:

$$M + Py = X, \tag{5.9}$$

$$\text{or} \quad EI\frac{\mathrm{d}^2y}{\mathrm{d}x^2} + Py = X. \tag{5.10}$$

This equation again gives rise to sinusoidal deflexions:

$$y = A\cos\alpha x + B\sin\alpha x + \frac{X}{P}, \tag{5.11}$$

from which

$$\frac{\mathrm{d}y}{\mathrm{d}x} = -\alpha A\sin\alpha x + \alpha B\cos\alpha x. \tag{5.12}$$

At the origin $x = 0$, $y = 0$ and $\mathrm{d}y/\mathrm{d}x = 0$, so that $A = -X/P$ and $B = 0$, and the deflexion y is given by

$$y = \frac{X}{P}(1 - \cos\alpha x). \tag{5.13}$$

Since the deflexion is zero at $x = l$,

$$\frac{X}{P}(1 - \cos\alpha l) = 0. \tag{5.14}$$

As before, the value of X cannot be zero, since a deflexion is assumed to have occurred, so that

$$\alpha l = 0,\ 2\pi,\ 4\pi, \ldots \tag{5.15}$$

and the Euler buckling load is

$$P_{\mathrm{e}} = 4\pi^2\frac{EI}{l^2}, \tag{5.16}$$

four times the value for the corresponding pin-ended column. As for the previous example, there are higher buckling modes which in practice are unobtainable.

[Note that for cos $\alpha l = 1$, then sin $\alpha l = 0$, and the apparently extra boundary condition of zero slope at $x = l$ is

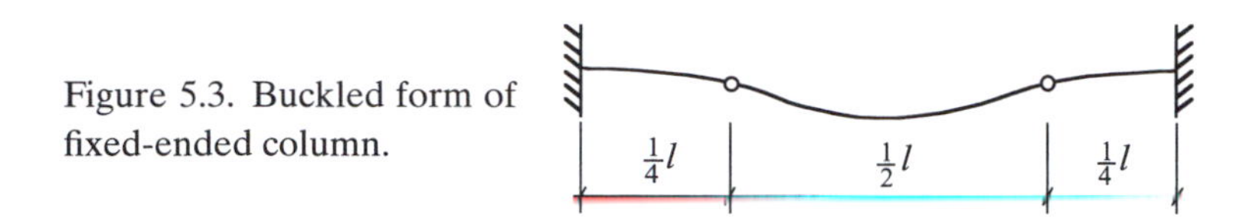

Figure 5.3. Buckled form of fixed-ended column.

automatically satisfied; four boundary conditions were specified, and only three unknown constants (A, B and X) entered the analysis. In fact, symmetry was assumed in fig. 5.2(a) – without this assumption, the two end moments should have been specified as X_1 and X_2.]

It was not really necessary to have pursued the analysis from eq. (5.9) onwards. The second-order equation must of necessity lead to a harmonic solution, and fig. 5.3 can be drawn at once as a (cosine) curve satisfying the zero-slope condition at the ends. Inflexion points occur at the quarter points, and the central portion of length $\frac{1}{2}l$ is the pin-ended column which was analysed to have the buckling load of eq. (5.7), in which l must now be replaced by $\frac{1}{2}l$; eq. (5.16) results.

The effective length of a fixed-ended column is thus half the length of the equivalent pin-ended column. Similarly, the effective length of a column with one end fixed and the other totally unrestrained (a flagpole) is $2l$. The intermediate case of a column with one end fixed in position and direction and the other end pinned is shown in fig. 5.4. The differential equation of bending may be written as:

$$EI\frac{d^2y}{dx^2} + Py = X - \frac{X}{l}x. \tag{5.17}$$

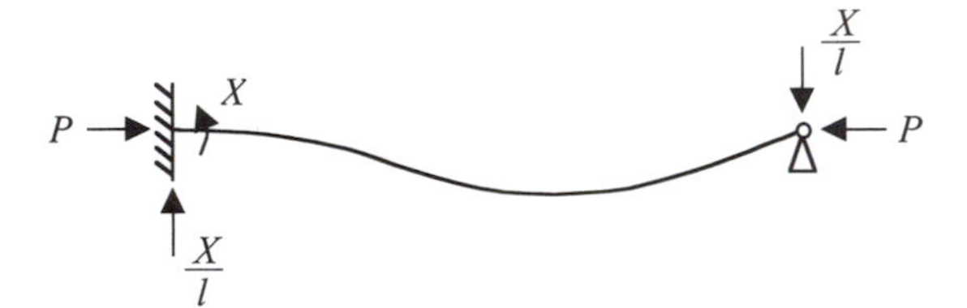

Figure 5.4. Buckling of a fixed/pinned column.

There are two constants of integration and the unknown bending moment X; use of the three boundary conditions leads to the buckling equation

$$\tan \alpha l = \alpha l, \tag{5.18}$$

of which the lowest root (other than zero) is $\alpha l = 4.493$. The buckling load is thus found to be

$$P_{\mathrm{e}} = \frac{20.19EI}{l^2} = \frac{\pi^2 EI}{(0.699l)^2}. \tag{5.19}$$

The propped cantilever in fig. 5.4 has an effective length against buckling of approximately $0.7l$.

The differential equations expressing buckling are of the form of eq. (5.2); eqs (5.10) and (5.17) have terms on the right-hand side representing the particular problem being studied. For all cases in which EI is constant, the basic equation may be written as

$$EI\frac{\mathrm{d}^4 y}{\mathrm{d}x^4} + P\frac{\mathrm{d}^2 y}{\mathrm{d}x^2} = 0, \tag{5.20}$$

and this holds for any segments between applied loads; the preliminary statical analysis (involving unknown redundant quantities) need not be carried out. Equation (5.20) solves to give

$$y = A\cos\alpha x + B\sin\alpha x + Cx + D, \qquad (5.21)$$

and the four constants of integration may be determined from the boundary conditions. (For the case of a pinned end, $d^2y/dx^2 = 0$.)

An approach to the design of practical columns may be made by the engineer assigning (with help from the building codes) an effective length to any particular member in a building frame. The effective length depends on its end connexions to other members; once chosen, a basic buckling strength may be calculated from eq. (5.7). This basic strength must, however, be modified to allow for other practical considerations.

5.2 Practical behaviour

The simple analysis given above indicates that, when a pin-ended column buckles, it can develop unrestrained lateral deflexions at a constant value of the (Euler) load; a similar conclusion holds for columns with other end conditions. This theoretical behaviour results from the approximation to the expression for curvature [eq. (3.12)] in which the term $(\mathrm{d}y/\mathrm{d}x)^2$ was neglected compared with unity. If the full expression is written (leading to highly non-linear equations), the load/lateral deflexion curve rises gently – the axial load must be progressively slightly increased to maintain increased deflexions, as shown schematically in fig. 5.5. If the central

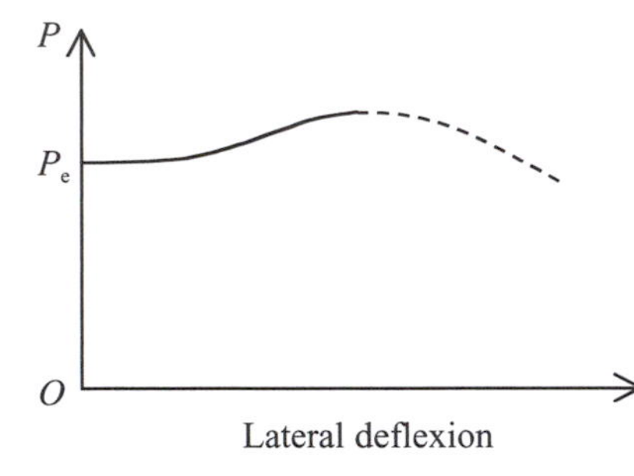

Figure 5.5. Post-buckled load/deflexion curve (schematic).

deflexion of the pin-ended column is a, then the magnitude of the bending moment there is Pa. Even at a constant value of P, the bending moment increases with the value of a, and at some stage becomes large enough to cause yield of the material, which in turn leads to rapidly increasing deflexions and perhaps catastrophic collapse, indicated by the dotted curve in fig. 5.5.

A practical design of column attempts to predict, and prevent, runaway behaviour, and is based on the analysis of the real members used in construction, notably the steel columns used in frames. No member is absolutely straight, and the introduction of an initial imperfection into the analysis leads to a more realistic approach to design. At the same time, the solution does not involve the eigenvalue property of the differential equation (to which there is in fact no solution until the critical load is reached). An initially bowed column may be represented by the curve

$$y_o = a_o \sin \frac{\pi x}{l}, \tag{5.22}$$

in which a_o is the central out-of-straightness of the member. Equation (5.22) could be regarded as the first term in a Fourier series representation of the initial shape; as will be seen, this term becomes dominant in the analysis.

When the axial load P is applied to the real pin-ended column, the resulting bending moment produces a *change* of curvature; that is, eq. (5.2) becomes

$$EI\left(\frac{d^2y}{dx^2} - \frac{d^2y_o}{dx^2}\right) + Py = 0, \tag{5.23}$$

that is,

$$\begin{aligned} EI\frac{d^2y}{dx^2} + Py &= -\frac{\pi^2 EI}{l^2}a_o \sin\frac{\pi x}{l} \\ &= -P_e a_o \sin\frac{\pi x}{l}. \end{aligned} \tag{5.24}$$

The general solution to this equation is

$$y = A\cos\alpha x + B\sin\alpha x + a_o\left(\frac{P_e}{P_e - P}\right)\sin\frac{\pi x}{l}. \tag{5.25}$$

The condition $x = 0$, $y = 0$ gives $A = 0$; and $x = l$, $y = 0$ gives

$$B\sin\alpha l = 0. \tag{5.26}$$

This is eq. (5.5), but there is now no bar to the solution $B = 0$, and the deflected form of the column is

$$y = a_o\left(\frac{P_e}{P_e - P}\right)\sin\frac{\pi x}{l}. \tag{5.27}$$

It is convenient to replace the second moment of area I by the expression $I = Ar^2$, where A is the area and r the radius of gyration of the cross-section, and to express the results in terms of a stress $\sigma = P/A$. Thus eq. (5.7) becomes

$$\sigma_{\mathrm{e}} = \pi^2 \frac{E}{(l/r)^2}, \tag{5.28}$$

where l/r is known as the *slenderness ratio*, and eq. (5.27) becomes

$$y = a_{\mathrm{o}} \left(\frac{\sigma_{\mathrm{e}}}{\sigma_{\mathrm{e}} - \sigma} \right) \sin \frac{\pi x}{l}; \tag{5.29}$$

the central deflexion is

$$a = a_{\mathrm{o}} \left(\frac{\sigma_{\mathrm{e}}}{\sigma_{\mathrm{e}} - \sigma} \right). \tag{5.30}$$

As the value of σ approaches σ_{e}, so the lateral deflexion increases very rapidly; the first term in a Fourier representation of the initial shape becomes dominant.

The total maximum compressive stress at mid-height of the column consists of the axial stress σ together with a bending stress due to the moment Pa, that is

$$\sigma + \frac{Pa_{\mathrm{o}}c}{I} \left(\frac{\sigma_{\mathrm{e}}}{\sigma_{\mathrm{e}} - \sigma} \right), \tag{5.31}$$

where c is the distance from the neutral axis of the section to the outermost compression fibre. Setting $I = Ar^2$ again, and assuming that the critical condition for the column occurs

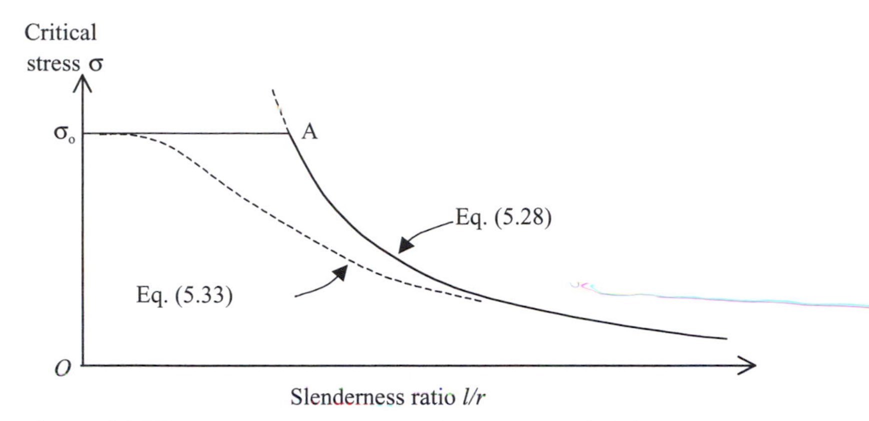

Figure 5.6. The Perry–Robertson buckling curve, eq. (5.33).

when the total compressive stress reaches the yield limit σ_o, then

$$\sigma_o = \sigma\left(1 + \frac{a_o c}{r^2}.\frac{\sigma_e}{\sigma_e - \sigma}\right). \tag{5.32}$$

This is a quadratic for the value of the critical stress σ, that is

$$\sigma^2 - \sigma\left[\sigma_o + (1 + \eta)\,\sigma_e\right] + \sigma_o\sigma_e = 0, \tag{5.33}$$

where $\eta = a_o c/r^2$. Equation (5.33) forms the basis of the rules for column design given in many codes of practice, and is shown in fig. 5.6. The value of η is assumed empirically by the codes (and is usually taken to be proportional to the slenderness ratio of the column); some discussion and a calculation are given in Appendix C.

Thus the practical design of columns involves major empirical elements, as well as some (perhaps unnoticed) structural assumptions. One of these assumptions is that column loads are axial, whereas real end connexions inevitably introduce some eccentricity of loading. In fact the response of a column to eccentric loading is very similar to that of an axially loaded but initially crooked member, and the empirical constant η does duty to cover both types of imperfection. Another major consideration, which requires some judgement on the part of the designer (aided by the design codes) is the concept of effective length – to cover the wide range of practical end connexions of a column in a building frame (or of a compressive member in a latticed truss).

The practical rules tend to be conservative, but not too wasteful. Once the end conditions of a particular member have been determined, and the working loads evaluated, then the rules lead to designs which are generally safe. In case of doubt, the size of a member may be increased to improve its buckling characteristics, often without any necessary penalty in weight or cost.

The determination of the end conditions of a compressive member in a hyperstatic structure is difficult. The problem does not arise for a statically determinate structure – for example, a pin-jointed latticed truss. If a compressive member in the determinate truss reaches its buckling limit [given by the solution of eq. (5.33)], then yield occurs accompanied by increasing lateral deflexions. The load/lateral deflexion curve for the member is shown schematically in fig. 5.7 (cf fig. 5.5), and the falling off of the load implies catastrophic collapse for the whole structure, albeit (for a structure with properly designed members) at the factored design collapse load. If a

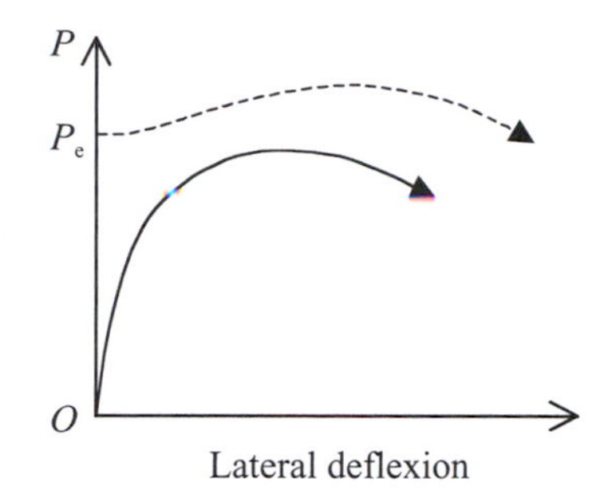

Figure 5.7. Practical load/deflexion curve (cf. fig. 5.5).

(ductile) tension member reaches its limiting load before any such buckling occurs, then overall collapse again occurs at the design load, but now as a quasi-static rather than a catastrophic process.

Behaviour is different for the hyperstatic structure. It was seen in Chapter 2 that the very simple braced square truss can accept an initial state of self-stress; if one of the diagonal members in fig. 2.4 were slightly oversize, then both diagonals would start from a state of initial compression. On the application of load, member BD may buckle when the external load is below its design value, as shown in fig. 2.7. If the initial loading point were at Y in this figure, then the member BD reaches its compressive limit of value C [now interpreted as the buckling load determined by eq. (5.33)], while member AC carries a load below its limiting tensile value. Figure 2.7 indicates that, in theory, the external load may continue to be increased until the tensile member reaches the limit T, but this conclusion must be examined in the light of the non-ductile nature of buckling of the compressive member. (Indeed, for any case in which the numerical value of C is less than that of

T, fig. 2.7 indicates that an initially stress-free truss may show the same behaviour.)

For the sake of simple calculations, the diagonal members *AC* and *BD* were assumed to be the same; in any case, whatever their relative dimensions, it was seen that the (elastic) extension of the tensile member *AC* was equal to the shortening of the compressive member *BD*. In general, despite possible plastic yielding in some members and possible buckling in others, if sufficient members remain elastic so that the structure remains a structure, then the strains in *all* members must be of the order of elastic strains. In fig. 2.4 the shortening of the buckling member *BD* is contained by the elastic strain of *AC*. It is, however, not clear that the limiting buckling load (of numerical value *C*) can be maintained as the external load on the truss is increased. The central deflexion of the initially bowed member has already increased from its initial value [eq. (5.30)], and further increase in compressive strain engenders further plastic yielding at the critical section.

The practical solution for the design of members in order to avoid such potentially catastrophic behaviour lies in the choice of load factors. It was noted in Chapter 2 that a way of introducing a load factor was to design the structure to just carry the specified loading – but to construct it with members having strengths greater than those calculated. Thus for steel frames in bending, with no problems of stability, or for tension members in trusses, a factor of 1.75 might be appropriate – the sections used would be 75 per cent stronger than the theoretical minima. For compressive members designed against buckling, the factor should be higher.

The factor of 1.75 is empirical, but not arbitrary. It results from the analysis of design rules refined in building codes over a period of nearly a century; steel structures designed according to those rules have been satisfactory. The same codes incorporate implicitly a higher factor against buckling – there is some variation, but a figure of about 2.5 has appeared to be appropriate. Some support is given for this figure by the numerical analysis in Appendix C.

5.3 Other buckling phenomena

All discussion in this book has been confined to planar structures – frames and trusses – but a different kind of buckling can occur in the third dimension, not necessarily involving axial load. A floor beam in a steel frame, for example, has been considered to act in bending (implicitly about its strong axis), as shown in the propped cantilever in fig. 2.12. If such a beam is not restrained properly against lateral movement, then it is possible that under critical loading the beam will both twist and deflect sideways. The analysis of such lateral-torsional buckling is more complex than that for the pin-ended column under axial load, but initially perfect members can exhibit the same eigenvalue properties. For example, analysis of a thin deep beam with pinned ends under the action of pure bending (i.e. a constant bending moment M) leads to the equation:

$$C\frac{\mathrm{d}^2\phi}{\mathrm{d}x^2} + \frac{M^2}{EI}\phi = 0, \tag{5.34}$$

where ϕ is the angle of twist of the beam, EI is the flexural rigidity of the cross-section about the *weak* axis, and C is the torsional rigidity of the section (some care must be taken in the evaluation of the constant C). Equation (5.34) is exactly analogous to eq. (5.2), and, as before, admits the trivial solution $\phi = 0$. The first meaningful solution requires that

$$M_e = \frac{\pi}{l}\sqrt{EIC}, \tag{5.35}$$

and the lateral/torsional displacements can in theory be of any magnitude.

Other loading conditions and end restraints lead to correspondingly more complex differential equations, which may be of fourth rather than second order, and whose solutions may involve hyperbolic or Bessel functions. All solutions have the same property, however – they lead to a critical value of loading similar to the Euler P_e for the simple strut.

A column forming part of a continuous building frame is subject to both bending and axial load; exact theoretical solutions may be found for very simple cases, but in general recourse must be had to concepts such as effective length to make any progress with a practical design. Similarly, a bending moment distribution which varies over the length of a column can be replaced by an equivalent moment M, and an interactive formula may be devised for a column subject to both bending and axial load:

$$\left(\frac{M}{M_e}\right)^2 + \left(\frac{P}{P_e}\right) = 1, \tag{5.36}$$

where M_e is given by a formula such as that in eq. (5.35), and P_e is the usual load for axial buckling. Finally, eq. (5.36) may be modified to allow for initial practical imperfections, and to consider possible additional bending about the weak axis.

Remarkably, despite the numerous assumptions, approximations and reliance on empirical data, structural members designed in this way can be used reliably in practical construction.

APPENDIX A

Virtual work

The equation of virtual work has been used throughout this book; it is a useful tool for manual calculation, and it enables easy proofs of the plastic theorems of Chapter 2 and of the elastic principles discussed in Chapter 4. A full proof of the equation is perhaps best presented in terms of a general stress analysis, but the following two outlines, in terms of frames and of trusses, while slightly incomplete, indicate the necessary steps of the argument.

A.1 Structures in bending

An initially horizontal beam, shown in fig. A1, is acted on by a distributed load w (which need not be uniform). A short length $\mathrm{d}x$ is cut from the beam (fig. A2), and to maintain vertical equilibrium of the element, a shear force F must act on the vertical faces. For equilibrium, it is seen that

$$\frac{\mathrm{d}F}{\mathrm{d}x} = w. \tag{A1}$$

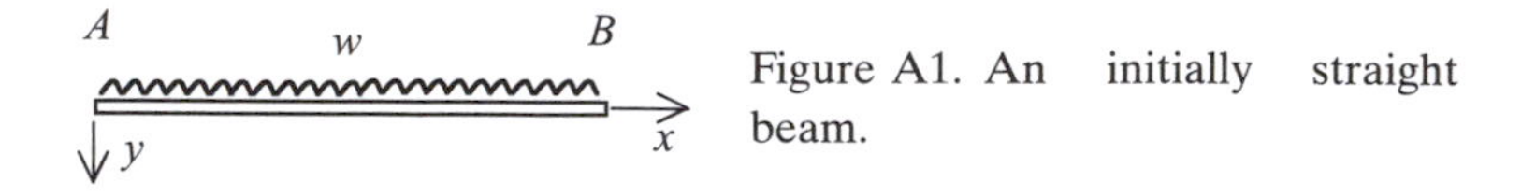

Figure A1. An initially straight beam.

Rotational equilibrium must also be satisfied, since the shear forces apply a couple of magnitude $F\mathrm{d}x$, and a bending moment M must also act on the vertical faces, to give

$$\frac{\mathrm{d}M}{\mathrm{d}x} = F. \tag{A2}$$

The shear force F may be eliminated between eqs (A1) and (A2), to give

$$\frac{\mathrm{d}^2 M}{\mathrm{d}x^2} = w. \tag{A3}$$

(The effect of shear force has not been discussed in this book, but it can be of paramount importance – for example, in the design of thin-walled box sections used as girders in bending.)

Equation (A3) is the basic equilibrium equation for a beam or for a straight member of a frame. The bending moments M are not necessarily produced in the member by

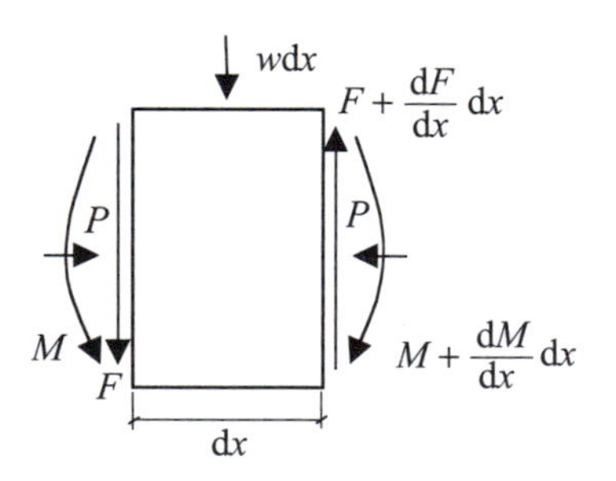

Figure A2. Equilibrium of beam element.

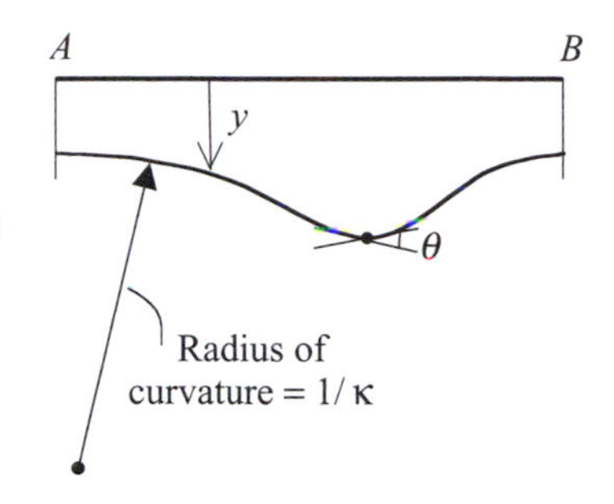

Figure A3. General displacements of beam.

the loading w. Indeed, eq. (A3) is of second order and, on double integration, the bending moments are determinable only in terms of the two constants of integration, and may be regarded as the superposition of free bending moments due to the loading and reactant self-stressing moments. An example is seen in the propped cantilever of figs 2.12(c) and (d), in which one of the constants of integration was determined from the condition that the bending moment at the right-hand prop is zero.

As a completely separate matter, the beam may be imagined to have displaced into the general position shown in fig. A3, under the action of some unspecified loading. The beam has deflected by an amount y at any particular section, and experiences a curvature κ at that section; also shown in the figure is a possible hinging discontinuity θ. As usual, all these quantities are small, and are related in the sense that

$$\text{deflexions } y \text{ are compatible with curvatures } \kappa \text{ and hinge discontinuities } \theta. \tag{A4}$$

More succinctly, the set (y, κ, θ) is compatible.

The equilibrium statement (A3) and the compatibility statement (A4) form the two unrelated components of the equation of virtual work.

Both sides of eq. (A3) may be multiplied by y to give

$$wy = \frac{\mathrm{d}^2 M}{\mathrm{d}x^2} y, \tag{A5}$$

where y is identified in due course as a displacement, but for the moment is assumed only to have those necessary requirements of continuity and differentiability necessary for the integration of eq. (A5). Indeed, on writing

$$\int wy \, \mathrm{d}x = \int \frac{\mathrm{d}^2 M}{\mathrm{d}x^2} y \, \mathrm{d}x \tag{A6}$$

and integrating twice by parts, it will be seen that

$$\int wy \, \mathrm{d}x = \int M \frac{\mathrm{d}^2 y}{\mathrm{d}x^2} \, \mathrm{d}x + \left[y \frac{\mathrm{d}M}{\mathrm{d}x} - M \frac{\mathrm{d}y}{\mathrm{d}x} \right], \tag{A7}$$

where the expression in square brackets must be evaluated at the limits of integration. If now:

(i) the equilibrium set (w, M) satisfies the loading boundary conditions for a given beam or frame;
(ii) the function y represents an imposed set of displacements satisfying the displacement boundary conditions, so that (y, κ) is a compatible set [$\kappa = \mathrm{d}^2 y/\mathrm{d}x^2$ from eq. (3.12)]; and
(iii) the integration extends over the whole structure,

then, from (iii), the term in square brackets in eq. (A7) must be evaluated at the external ends of the members of the system.

For the three simple conditions of a free, pinned, or clamped end, the following relations hold:

$$\left.\begin{array}{lll}\text{Free end:} & \left.\begin{array}{r} M=0 \\ F=\frac{\mathrm{d}M}{\mathrm{d}x}=0\end{array}\right\} \\ \text{Pinned end:} & \left.\begin{array}{r} M=0 \\ y=0\end{array}\right\} \\ \text{Clamped end:} & \left.\begin{array}{r} y=0 \\ \frac{\mathrm{d}y}{\mathrm{d}x}=0\end{array}\right\}\end{array}\right\} \tag{A8}$$

For all these end conditions, the term in square brackets in eq. (A7) vanishes, so that

$$\int wy\,\mathrm{d}x = \int M\kappa\,\mathrm{d}x. \tag{A9}$$

For other end conditions (e.g. an elastic support), eq. (A9) is valid provided that the reactions are introduced into the equation as external loads. Similarly, due attention must be given to the internal connexions between the members of a frame.

Equation (A9) is the basic virtual-work equation for structures in bending, relating an equilibrium set (w, M) with a compatible set (y, κ). It must be emphasized that there is no necessary connexion between the two sets. The bending moments M can have the actual values under the external loads w, or could be any equilibrium set satisfying eq. (A3). Similarly, the set (y, κ) may represent the actual deformed state of the frame, or could represent any compatible set of imposed displacements.

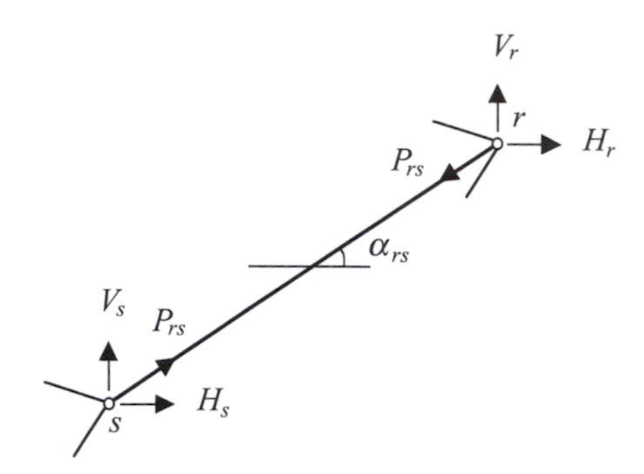

Figure A4. Typical bar rs of a pin-jointed truss.

Equation (A9) may be expanded to allow both for concentrated loads and for the effects of sudden changes in curvature (i.e. hinge discontinuities); in the expanded form

$$\sum W_i y_i + \int wy \, \mathrm{d}x = \sum M_k \theta_k + \int M\kappa \, \mathrm{d}x. \qquad \text{(A10)}$$

On the left-hand side, the summation includes all concentrated loads W_i, and the integral extends over all other loads; on the right-hand side, the summation includes all hinge discontinuities θ_k (where the corresponding values of bending moments are M_k), and the integration extends over all the rest of the frame.

A.2 Trusses

Figure A4 shows a bar *rs* of a plane pin-jointed truss; other bars also meet at the two joints. A load W_r acts at joint r, and is represented by its horizontal and vertical components H_r and V_r. Similarly, a load W_s acts at joint s. The bar is inclined

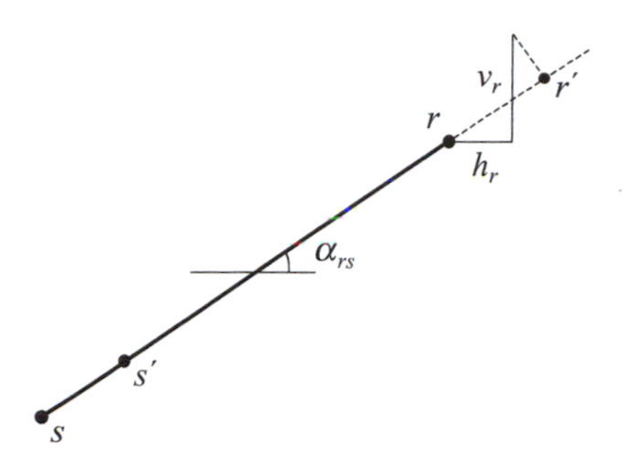

Figure A5. Displacements of bar rs.

at an angle α_{rs} to the horizontal. The following two statical equations may be written for joint r:

$$\left.\begin{aligned} H_r &= \sum_{bars} P_{rs} \cos \alpha_{rs} \\ V_r &= \sum_{bars} P_{rs} \sin \alpha_{rs}, \end{aligned}\right\} \tag{A11}$$

where the summation is carried out over all the bars meeting at joint r. The first of eqs (A11) may be multiplied by a quantity h_r and the second by v_r; these two quantities may be thought of as numbers, but will be identified as components of a virtual (and independent) displacement of joint r. The two modified equations are then added to give

$$H_r h_r + V_r v_r = \sum_{bars} \{P_{rs} (h_r \cos \alpha_{rs} + v_r \sin \alpha_{rs})\}. \tag{A12}$$

Figure A5 identifies the displacement in simple brackets on the right-hand side of eq. (A12) as the extension rr' of the bar rs due to the imposed displacements h_r and v_r.

Equation (A12) is written for joint r; similar equations may be written for all the other joints of the truss. If now all these equations are summed, the bar force P_{rs} appears just twice – once in eq. (A12), and once in the similar equation for joint s. The displacement term in brackets for joint s is the distance ss' marked in fig. A5; moreover, it is seen in fig. A4 that the bar force P_{rs} occurs with a negative sign when resolution of forces is carried out at joint s. Thus in the summation of all the equations similar to eq. (A12), the bar force P_{rs} is multiplied by $(rr' - ss')$ – that is, by the extension e_{rs} of the bar rs. Finally, then,

$$\sum_{joints} (H_r h_r + V_r v_r) = \sum_{joints} \left\{ \sum_{bars} (P_{rs} e_{rs}) \right\}. \qquad \text{(A13)}$$

The left-hand side of the equation is the vector product of a load W having components H and V with a displacement Δ having components h and v, so that, dropping suffices,

$$\sum W \cdot \Delta = \sum Pe. \qquad \text{(A14)}$$

This is the equation of virtual work for trusses, and it is evident from the derivation that there is no necessary connexion between the equilibrium statement (W, P) and the compatibility statement (Δ, e).

APPENDIX B

The plastic theorems

The proofs of the plastic theorems follow from applications of the equation of virtual work, and can be given most easily in terms of the simple framed structure. The rectangular portal frame in fig. B1(a) (which is meant to represent far more complex structures) is acted upon by loads W; all loads have the same multiplying load factor λ. The value λ_c at collapse is sought. As discussed in Chapter 2, the three master statements of structural theory may be written as:

Equilibrium:	Internal bending moments M in the frame are in equilibrium with the external loads W.
Yield:	The values of M are less than, or at most equal to, the value of the full plastic moment M_p.
Mechanism:	There is an arrangement of plastic hinges which will permit deformation of the frame.

Figure B1(b) shows a mechanism of deformation, and the rotations θ are compatible with displacements Δ of the loading

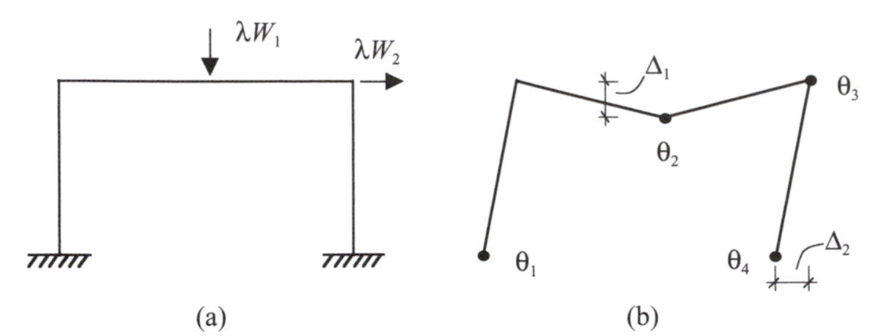

Figure B1. Schematic representation of a framed structure.

points. The internal work dissipated at a plastic hinge is $M_p|\theta|$, which is always positive; the value of M_p may vary from point to point around the frame.

Uniqueness

It will be supposed that, for a given loading on the frame, there are two different collapse mechanisms formed at different load factors λ^* and λ^{**}. For the first mechanism the collapse bending moments around the frame are given by a distribution M^*, where the equilibrium equations are satisfied and $|M^*| \leq M_p$; the mechanism of collapse is (Δ^*, θ^*). A similar statement may be made for collapse at the load factor λ^{**}, so that

$$\left.\begin{array}{l} A:(\lambda^* W, M^*) \text{ satisfy the equilibrium and yield conditions.} \\ B:(\lambda^{**} W, M^{**}) \text{ satisfy the equilibrium and yield conditions.} \\ C:(\Delta^*, \theta^*) \text{ describes a mode of plastic deformation.} \\ D:(\Delta^{**}, \theta^{**}) \text{ describes a mode of plastic deformation.} \end{array}\right\} \quad \text{(B1)}$$

The collapse equation for the first mechanism may be written by combining statements A and C in the equation of virtual work:

$$\sum \lambda^* W \Delta^* = \sum M^* \theta^*, \tag{B2}$$

The value of $|M^*|$ at each hinge position is equal to M_p, so that the collapse load factor λ^* is given by

$$\lambda^* \sum W \Delta^* = \sum M_p \left|\theta^*\right|. \tag{B3}$$

Statements *B* and *C* in eq. (B1) can also be combined by the equation of virtual work:

$$\lambda^{**} \sum W \Delta^* = \sum M^{**} \theta^*. \tag{B4}$$

The bending moments M^{**} satisfy the yield condition; that is, if mechanisms θ^* and θ^{**} have a common hinge, then $|M^{**}| = M_p$ at that hinge, but otherwise $|M^{**}| < M_p$ at the hinge points of the mechanism θ^*. Thus, in eq. (B4),

$$\sum M^{**} \theta^* \leq \sum M_p \left|\theta^*\right|, \tag{B5}$$

so that

$$\lambda^{**} \sum W \Delta^* \leq \sum M_p \left|\theta^*\right|. \tag{B6}$$

Comparing eq. (B3) with inequality (B6),

$$\lambda^{**} \leq \lambda^*. \tag{B7}$$

Statement D in eq. (B1) has not been used, and if the arguments are repeated with statement D instead of statement C, then

$$\lambda^* \leq \lambda^{**}. \tag{B8}$$

Thus λ^* and λ^{**} have the same value – namely, the collapse value λ_c. The proof has shown only that the load factor at collapse is unique. Nothing has been proved about the mode of deformation; indeed, it is possible for different modes to exist at the same value of collapse load factor.

The upper bound theorem (the unsafe theorem)

The theorem states that if a plastic mode of deformation is assumed, and the work done by the external loads is equated to the internal work dissipated, then the resulting load factor λ' is always greater than, or at best equal to, the true load factor λ_c. The following statements are used:

$$\left.\begin{aligned} &E: (\lambda_c W, M_c) \text{ is the actual collapse distribution.} \\ &F: (\Delta', \theta') \text{ is the assumed collapse mechanism.} \end{aligned}\right\} \tag{B9}$$

The work equation for the assumed collapse mechanism is

$$\lambda' \sum W\Delta' = \sum M_p \left|\theta'\right|. \tag{B10}$$

Statements E and F of eq. (B9) combine to give

$$\lambda_c \sum W\Delta' = \sum M_c \theta'. \tag{B11}$$

Now $|M_c| \leq M_p$, so that, following the previous arguments,

$$\lambda_c \sum W\Delta' \leq M_p \left|\theta'\right|, \tag{B12}$$

and comparison of eqs (B10) and (B12) shows that

$$\lambda_c \leq \lambda'. \tag{B13}$$

The lower bound theorem (the safe theorem)

The theorem states that if a set of bending moments can be found that satisfies the equilibrium and yield conditions at a yield factor λ'', then λ'' is always less than, or at best equal to, the true load factor λ_c. The following statements are used:

$$\left.\begin{array}{l} G:(\lambda''W, M'')\text{ represents a set of bending moments} \\ \quad\text{satisfying the equilibrium and yield conditions.} \\ H:(\lambda_c W, M_c)\text{ is the actual collapse distribution.} \\ J:(\Delta_c, \theta_c)\text{ is the actual collapse mechanism.} \end{array}\right\} \tag{B14}$$

Statements H and J give

$$\lambda_c \sum W\Delta_c = \sum M_c\theta_c = M_p \left|\theta_c\right|, \tag{B15}$$

while statements G and J give

$$\lambda'' \sum W\Delta_c = \sum M''\theta_c \leq M_p \left|\theta_c\right| \tag{B16}$$

as before. Hence

$$\lambda'' \leq \lambda_c. \tag{B17}$$

The results may be displayed compactly [cf eq. (2.10) in Chapter 2)] as

$$\lambda = \lambda_c \left\{ \begin{array}{ll} \left. \begin{array}{l} \text{Equilibrium condition} \\ \text{Yield condition} \end{array} \right\} & \lambda \leq \lambda_c \\ \text{Mechanism condition} & \lambda \geq \lambda_c \end{array} \right\} \qquad \text{(B18)}$$

APPENDIX C

Buckling calculations

The empirical constant η was introduced into eq. (5.33) to allow for an initial central bow a_o in a pin-ended column. The value of η is given by

$$\eta = \frac{a_o c}{r^2} = \frac{a_o}{l} \cdot \frac{c}{r}\left(\frac{l}{r}\right) = k\left(\frac{l}{r}\right). \tag{C1}$$

The values of c, the distance of the extreme compression fibre from the neutral axis, and of r, the radius of gyration, are physical properties of any cross-section – for a rectangular cross-section, for example, $c/r = \sqrt{3}$. For a range of Universal Column sections, the value of c/r is very close to 2.

Empirical values of k in eq. (C1) given in design codes are in the range 10^{-3} to 3×10^{-3}; using the value $c/r = 2$, then eq. (C1) shows that values of a_o/l lie in the range 0.5×10^{-3} to 1.5×10^{-3}. This range is used in the numerical calculations given below.

Figure 5.6 indicates that the greatest difference between perfect and actual behaviour occurs when the theoretical buckling stress σ_e [eq. (5.28)] is equal to the yield stress σ_o, shown as A in fig. 5.6. For steel with $E = 210 \times 10^3$ N/mm^2 and $\sigma_o = 250$ N/mm^2, eq. (5.26) shows that

$$\sigma_o = \sigma_e = \frac{(\pi^2)(210 \times 10^3)}{(l/r)^2} = 250, \qquad \text{(C2)}$$

from which the slenderness ratio at point A in fig. 5.4 is determined as $l/r = 91$.

Setting $\sigma_o = \sigma_e$ in eq. (5.33), the critical stress σ is given by the solution of

$$\left(\frac{\sigma}{\sigma_o}\right)^2 - (2 + \eta)\left(\frac{\sigma}{\sigma_o}\right) + 1 = 0, \qquad \text{(C3)}$$

and for $l/r = 91$, the empirical range for η is 91×10^{-3} to 273×10^{-3}. The corresponding values of (σ/σ_o) from eq. (C3) are 0.74 and 0.60.

Equation (5.30) shows how the deflexions increase from a_o to a as σ approaches σ_e; for the range a_o/l of 0.5 to 1.5×10^{-3} given previously, the corresponding range for a/l at the critical stress is (0.5)/(0.26) to (1.5)/(0.40); that is, 1.92 to 3.75×10^{-3}. The resultant axial shortening (i.e. the approach of the ends of the column) is proportional to the square of such small lateral displacements, and is negligible compared with the elastic axial shortening resulting from direct axial compression. Thus at the critical condition for the column being studied, the elastic compressive strain lies in the range 0.74 to 0.60 of the yield strain of the material.

For the simple truss which has been studied throughout this book it was seen (fig. 2.7) that the extension of the tension member AC was equal to the compression of the diagonal BD;

if diagonal BD suffers a compressive strain in the range 0.74 to 0.60 of the yield strain, this is exactly the tensile strain suffered by the diagonal AC. However, the truss is now in a critical state; a certain value of external load has caused the compression member to reach its critical condition – yield is just occurring. An attempt to increase the external load further causes yield to spread rapidly through the section; the lateral deflexion increases markedly, and the axial compressive strain increases, so that the tensile strain in AC also increases until the limiting tensile load is reached. However, this entire process may be unstable – instead of a quasi-static collapse of the truss, failure may be sudden.

The example discussed here is highly artificial, but a general conclusion may be drawn. In a hyperstatic truss which is not on the point of collapse, at least one of the members is in an elastic state, and the strain in that member is below the elastic limit. The strains in all the other members must be of the same elastic order, even if one or more of the members has yielded; the members of the truss must still fit together, and any plastic deformation which may have occurred is contained in magnitude by the strain of the last unyielded member. If, however, the last member to fail does so by buckling, it may prove impossible to achieve the required design load, and, moreover, failure may occur catastrophically.

Buckling must, in fact, be avoided in practical construction. A simple way of doing this is to increase the strength of the compression members. For the simple square truss in fig. 2.7, it was shown that a critical compressive stress σ/σ_o lies in the range 0.74 to 0.60. If the relevant compression members were manufactured from material having a yield stress σ_0 in the range 1.3 to 1.7 times that of the tension members, the tension members would yield before a compression member buckled. Alternatively – and this is the device implicit in practical codes of practice – the safety factor against buckling can be set higher than the factor against tensile (or bending) yield. It was noted that one way of incorporating a load factor in design is to arrange for a structure to be just on the point of collapse under the specified values of load, and to actually build the structure stronger. Thus tensile or bending members in steel could be manufactured to be 1.75 times as strong as the theoretical minima. Using the range 0.74 to 0.60 achievable by compression members, the load factor corresponding to 1.75 would lie in the range 2.36 to 2.92, and compression members could be manufactured to have buckling strengths say 2.5 times the theoretical minima.

Index